02-ECO-1

WATER QUALITY INDICATORS: NUTRIENT IMPACTS ON CHLOROPHYLL OR ALGAE SPECIES COMPOSITION

Kenneth H. Reckhow, Ph.D.
Duke University

George B. Arhonditsis, Ph.D.
Melissa A. Kenney
Sandra J. McBride, Ph.D.
Duke University

Russell J. Gosnell, Ph.D.
North Carolina Central University

Craig A. Stow, Ph.D.
University of South Carolina, Columbia

Hans W. Paerl, Ph.D.
University of North Carolina, Chapel Hill

2006

WERF

The Water Environment Research Foundation, a not-for-profit organization, funds and manages water quality research for its subscribers through a diverse public-private partnership between municipal utilities, corporations, academia, industry, and the federal government. WERF subscribers include municipal and regional water and wastewater utilities, industrial corporations, environmental engineering firms, and others that share a commitment to cost-effective water quality solutions. WERF is dedicated to advancing science and technology addressing water quality issues as they impact water resources, the atmosphere, the lands, and quality of life.

For more information, contact:
Water Environment Research Foundation
635 Slaters Lane, Suite 300
Alexandria, VA 22314-1177
Tel: (703) 684-2470   Fax: (703) 299-0742
www.werf.org   werf@werf.org

This report was co-published by the following organization. For nonsubscriber sales information, contact:

IWA Publishing
Alliance House, 12 Caxton Street
London SW1H 0QS, United Kingdom
Tel: +44 (0) 20 7654 5500
Fax: +44 (0) 20 7654 5555
www.iwapublishing.com
publications@iwap.co.uk

Library of Congress Catalog Card Number: 2005932225

IWAP ISBN 978-1-84339-747-2

This report was prepared by the organization(s) named below as an account of work sponsored by the Water Environment Research Foundation (WERF). Neither WERF, members of WERF, the organization(s) named below, nor any person acting on their behalf: (a) makes any warranty, express or implied, with respect to the use of any information, apparatus, method, or process disclosed in this report or that such use may not infringe on privately owned rights; or (b) assumes any liabilities with respect to the use of, or for damages resulting from the use of, any information, apparatus, method, or process disclosed in this report.

Duke University, University of North Carolina, University of South Carolina, North Carolina Central University

This document was reviewed by a panel of independent experts selected by WERF. Mention of trade names or commercial products does not constitute WERF nor EPA endorsement or recommendations for use. Similarly, omission of products or trade names indicates nothing concerning WERF's nor EPA's positions regarding product effectiveness or applicability.

The research on which this report is based was funded, in part, by the United States Environmental Protection Agency through Cooperative Agreement No. CR-827345-01 with the Water Environment Research Foundation (WERF). Unless an EPA logo appears on the cover, this report is a publication of WERF, not EPA. Funds awarded under the Cooperative Agreement cited above were not used for editorial services, reproduction, printing, or distribution.

ACKNOWLEDGMENTS

This research was funded by the Water Environment Research Foundation through the University of North Carolina Water Resources Research Institute. Thanks are extended to Jordan Williams and Tammy Hill for their assistance with the graphic art and document formatting for this report. Lauren Hauser, Jenny Tribo, Chris Wu, Karen Elcock, Linda Caldwell, Matthew Barkley, and Emily Estes completed case studies on predictive nutrient criteria for their Masters of Environmental Management Project at Duke University. Z. Xiao completed research on robust estimation for multivariate outlier detection for his Masters of Science in mathematics at North Carolina Central University. Finally, Modupe Banjoko and Marissa Massey completed research projects related to predictive nutrient criteria as part of an independent study project at North Carolina Central University.

Report Preparation

Principal Investigator:
Kenneth H. Reckhow, Ph.D.
Duke University

Project Team:
George B. Arhonditsis, Ph.D.
Melissa A. Kenney
Sandra J. McBride, Ph.D.
Duke University

Russell J. Gosnell, Ph.D.
North Carolina Central University

Craig A. Stow, Ph.D.
University of South Carolina, Columbia

Hans W. Paerl, Ph.D.
University of North Carolina, Chapel Hill

Project Subcommittee

Norman E. LeBlanc
Hampton Roads Sanitation District

Stephen Bainter
U.S. Environmental Protection Agency

Larry W. Haas, Ph.D.
Virginia Institute of Marine Science, College of William and Mary

Michael C. Harrass, Ph.D.
US Borax

Tom Swan, Ph.D.
Truckee Meadows Water Reclamation Facility

Jay H. Sauber
North Carolina Division of Water Quality

Andrea C. Rex, Ph.D.
David Taylor, Ph.D.
Massachusetts Water Resources Authority

Water Environment Research Foundation Staff

Director of Research: Daniel M. Woltering, Ph.D.
Program Director Margaret Stewart

ABSTRACT AND BENEFITS

Abstract:

The objective of this project was to develop an alternative and more appropriate method to select nutrient criteria. To accomplish this objective the research team conducted a study in two major areas: 1) nutrient criteria that are linked to the designated use and 2) key aquatic ecological patterns relevant to developing predictive nutrient criteria.

The major focus of our research was to evaluate the scientific basis for nutrient criteria as predictors of designated uses. Because 100% attainment of a designated use us impractical for most waterbodies, the method that we developed acknowledges that policymakers need to assess what probability of nonattainment is acceptable to set appropriate criteria levels. To answer the research question, the team used a combination of methods that included: studying eutrophication theory from literature reviews, examining water quality data, collecting expert elicitations, and building structural equation models. With these tools, the team was able to study eutrophication impairment in lakes, reservoirs, estuaries, and coastal ecosystems.

The present research introduces an alternative method for developing nutrient criteria, that predictively links the criteria to the water quality goal. This method explicitly acknowledges the importance of connecting the elicited probabilities of use attainment with realistic ecological structures. Given the inherent policy tradeoff between maximizing environmental protection and minimizing total maximum daily load (TMDL) research and implementation costs, the project offers insights that are useful to decision makers when selecting appropriate water quality standards.

Benefits:

- Develops an alternative to select nutrient criteria that is statistically linked to the water quality goal.
- Creates a methodological tool that can quantify the relationship between the level (concentration) of the criterion and the probability of designated use attainment.
- Provides a method that can be generalized at a wider scale and can be applied to certain regions, political units, or broader landscape/climate/geological features by analyzing multiwaterbody cross-sectional datasets and eliciting multiple expert judgments.

Keywords: Designated use, water quality criteria, structural equation modeling, expert elicitation, phytoplankton dynamics

TABLE OF CONTENTS

LIST OF TABLES

LIST OF FIGURES

LIST OF ACRONYMS

DIN	dissolved inorganic nitrogen
DIP	dissolved inorganic phosphorus
DO	dissolved oxygen
NCDWQ	North Carolina Division of Water Quality
NOX	nitrogen oxide
PFG	phytoplankton functional group
RMSE	root mean squared error
SEM	structural equation modeling
SRP	soluble reactive phosphorus
TMDL	total maximum daily load
U.S. EPA	U.S. Environmental Protection Agency
WERF	Water Environment Research Foundation

EXECUTIVE SUMMARY

The objective of this project was to develop an alternative and more appropriate method to select nutrient criteria. The major focus of this research was to evaluate the scientific basis for nutrient criteria as a predictor of designated use attainment. To answer the research question, the research team used a combination of methods that included studying eutrophication theory from literature reviews, examining water quality data, collecting expert elicitations, and building structural equation models. With these tools, researchers may study eutrophication impairment in lakes, reservoirs, estuaries, and coastal ecosystems.

ES.1 Predictive Nutrient Criteria

In 1998, the President's Clean Water Action Plan directed the U.S. Environmental Protection Agency (U.S. EPA) to develop a national strategy for establishing nutrient criteria (U.S. EPA, 1998). The resultant multiyear study produced a set of documents and recommended criteria based on ecoregions and waterbody type.

In principle, standard setting should be viewed from the perspective of decision making under uncertainty, which inherently involves interplay between science and public opinion. The determination of designated uses reflects public values, both in the statements in the Clean Water Act and in the waterbody-specific statement of designated use. However, the means to protect these designated uses—through water quality criteria—is a choice left largely to the state. Therefore, a good candidate criterion is one that is easily and reliably measured and is a good indicator of designated use, a choice that should be dictated largely by science.

The determination of the criterion level associated with the attainment–nonattainment transition, however, requires the integration of science and values. Natural variability and scientific uncertainty in the relationship between the criterion and the designated use imply that selection of a criterion level with 100% assurance of use attainment is generally unrealistic. Accordingly, scientific uncertainty and attitude toward risk of nonattainment should be part of the criterion level decision. Therefore, the decision on criterion level might be addressed by answering the following question: Acknowledging that 100% attainment is impractical for most criteria, what probability (or, perhaps, what percentage of space–time) of nonattainment is acceptable? Analytically, this question may be answered by integrating the probability of use attainment, for a given criterion level, with a decision maker's utility function, which reflects water quality costs and benefits. The criterion level associated with the highest expected utility might then be chosen. Realistically, this decision analytic framework is prescriptive; it guides researchers toward what ought to be done, but it almost certainly exceeds what actually will be done.

The team developed a methodological framework that predicts probability of use attainment as a function of criterion level, and leaves the choice of the criterion level to policy makers. Once the procedure was developed, the major challenges in this work were the quantification of the narrative designated use statement and the selection of a modeling strategy that accurately describes the causal structures underlying the system ecological functions. The latter was effectively addressed by the application of structural equation modeling (SEM); here, the researchers illustrate some of the attractive features of this method that can be very useful for researchers when assessing eutrophic conditions and disentangling complex environmental management issues. The former was initially supported by a two-stage expert elicitation with a single expert and further fortified by the inclusion of multiple experts. The research team also employed proper procedures to combine expert judgments and gave special emphasis on the reduction of human errors caused by heuristics and elicitation bias by conducting consistency checks and articulating the questionnaires.

The team applied the proposed methodological framework on four single-waterbody (Lake Washington, Lake Mendota, San Francisco Bay, and the Neuse Estuary) and two multiwaterbody (North Carolina and Florida) datasets. This report includes three of these studies for illustrative purposes: Lake Washington, the Neuse River Estuary, and western North Carolina region.

ES.2 Ecological Patterns

Another part of this project was to use structural equation models to separate out the roles of nutrients in ecological patterns, which are the basis of nutrient criteria development. It was important to explore the role of nutrients in these processes so that one can discern the difference between the biological factors (grazing) and nonbiological factors (stratification/salinity gradients) on ecological function. To accomplish this goal, the team formulated a simple conceptual model that considered the regulatory role of abiotic factors and biological interactions on epilimnetic phytoplankton dynamics and water clarity during the summer stratification period. Both eutrophic (Lake Mendota) and mesotrophic (Lake Washington) conditions were used to test the initial hypothesized model and assess the relative importance of several ecological processes (vertical mixing, intrusions of the hypolimnetic nutrient stock, herbivory) known to determine the levels of water quality variables of management interest (water clarity, cyanobacteria).

Additionally, the researchers introduced a Bayesian SEM approach to evaluate the relative importance of the physical environment (flow, salinity, temperature, and light availability) versus nutrient concentration (inorganic nitrogen and phosphate) effects on the spatiotemporal phytoplankton biomass and community composition patterns in an estuarine environment. Bayesian analysis is a statistical method that incorporates prior information with data to create probabilistic assessments. In this research project, a Bayesian hierarchical method was used to make the model identification restrictions stochastic, provide information on multimodality, and relax SEM's asymptotic theory. Because of the flexibility of the Bayesian approach, it is particularly useful in detangling complex ecological patterns.

More specifically, the team delineated the hydrological forcing *vis-à-vis* the nutrient role on phytoplankton growth and compositional alterations along the Neuse River Estuary. The research team therefore resolved the particularly problematic relationship between factors that often have a confounding effect on the waterbody attributes. At the same time, the team gained basic knowledge of the complex interactions that drive phytoplankton community dynamics in estuarine ecosystems, influencing their responses to nutrient input reductions. The results of these studies are presented in the appendixes.

ES.3 Contributions and Relevance

This project develops a methodological framework for setting water quality criteria from the prescriptive basis that criteria should be predictive of designated use and from the pragmatic basis that risk of nonattainment should be acknowledged, and, therefore, considered when setting a level or concentration. Furthermore, the selection of the acceptable probability, or criteria level, is a value judgment best left to policy makers and should result from a value tradeoff between maximizing environmental protection and minimizing TMDL costs.

ES.4 Resulting Publications

The following publications have been submitted or published from this Water Environment Research Foundation (WERF) research:

1. Reckhow, K.H.; Arhonditsis, G.B.; Kenney, M.A.; Hauser, L.; Tribo, J.; Wu, C.; Elcock, K. J.; Steinberg, L.J.; Stow, C.A.; McBride S.J. A Predictive Approach to Nutrient Criteria. *Environ. Sci. Technol.* **2005** *39*, 2913–2919.
2. Arhonditsis, G.B.; Stow, C.A.; Steinberg, L.J.; Kenney, M.A.; Lathrop, R.C.; McBride, S.J.; Reckhow, K.H. Exploring Ecological Patterns with Structural Equation Modelling and Bayesian Analysis. *Ecol. Model.* **2006** *192*, 385–409.
3. Arhonditsis, G.B.; Paerl, H.W.; Valdes, L.M.; Stow, C.A.; Steinberg, L.J.; Reckhow, K.H. Application of Bayesian Structural Equation Modeling for Exploring the Neuse River Estuary (NC, USA) Phytoplankton Dynamics. *Limnol. Oceanogr.,* submitted manuscript.
4. Arhonditsis, G.B.; Stow, C.A.; Paerl, H.W.; Valdes, L.M.; Steinberg, L.J.; Reckhow, K.H. Delineation of the Role of Nutrient Dynamics and Hydrologic Forcing on the Neuse River Estuary (NC, USA) Spatiotemporal Phytoplankton Patterns. *Ecosystems*, submitted manuscript.

CHAPTER 1.0

INTRODUCTION

This project developed an alternative method to select nutrient criteria. Because one of the major goals of criteria is to serve as a surrogate for the designated uses, a major focus of the research was to evaluate the scientific basis for nutrient criteria as a predictor of designated use attainment. To address the research question, the team used a combination of methods that included studying eutrophication theory from literature reviews, examining water quality data, collecting expert elicitations, and building structural equation models. With these tools, the research team was able to study eutrophication impairment in a variety of waterbody systems, including lakes, reservoirs, estuaries, and coastal ecosystems.

1.1 Overview of the Project

The U.S. Environmental Protection Agency (U.S. EPA) considers more than 25,000 of the waters in the United States impaired. Eutrophication is one of the top four sources of waterbody impairment (U.S. EPA, 2000b).Eutrophication, which is fueled by excess nitrogen and phosphorous, causes problems such as anoxia, noxious algal blooms, and fish kills. These eutrophication symptoms result in noncompliance of a waterbody's water quality standards.

Eutrophication, unlike other pollutants, manifests itself differently in different systems. For example, if the same amount of nitrogen and phosphorous was added to two lakes, the ecological response could be very different in each of the systems. As a result, eutrophication symptoms are not caused by predictable amount of nutrients. Therefore, there is a need to understand better the relationship between nutrients and algal response in aquatic ecosystems.

Despite the somewhat unpredictable response, it is important to protect waterbodies from excessive eutrophication-related impairments by setting appropriate nutrient criteria. There is debate about whether the most appropriate criteria are the causal water quality variables (nitrogen and phosphorous) or the response variables (such as chlorophyll *a*). Part of this question can be answered through a better understanding of the relationships between water quality variables and eutrophication-related water quality impairment.

The team conducted research to understand better how to develop a method that would statistically link water quality criteria and the designated use. Details on this project are presented in more detail in the following chapter.

CHAPTER 2.0

PREDICTIVE NUTRIENT CRITERIA

2.1 Overview of the Nutrient Criteria

Water quality standards are described in Sections 101(a) and 303(c) of the Clean Water Act, and are set using an antidegradation clause, a qualitative designated use, and a qualitative or quantitative criterion. The antidegradation clause is a narrative statement that states the designated use cannot be lessened because of degraded water quality conditions. The designated use is a narrative statement that describes the water quality goal. The criterion is a numeric measure or narrative statement that serves as a scientific surrogate for the designated use.

The designated use is important because it describes those water quality goals that have the most meaning to the waterbody users. At a minimum, Section 101(a) of the Clean Water Act requires that, where attainable, water quality must provide for the protection of propagation of fish, shellfish, and wildlife, and recreation in and on the water (also referred to as fishable/swimmable). States are allowed to set designated uses that are more specific or more restrictive than the fishable/swimmable designation (e.g., cold water habitats and drinking water supplies).

Designated uses are narrative descriptions and cannot be directly assessed with a single water quality variable, therefore it is difficult to determine whether a waterbody is meeting its designated use. Ideally, one could use a scientific test kit to say whether a waterbody is meeting or is not meeting the goal of fishable/swimmable. Unfortunately, such methods do not exist; instead, one must use measurable water quality criteria as proxies for assessing attainment of a designated use. Hence, water quality standards are not enforced by the direct assessment of the designated use; they are assessed through the water quality criterion. Because an indirect measure of the water quality goal is used to assess attainment, it is prudent to question whether this criterion is appropriate. If the relationship between the designated use and the criterion was never rigorously tested, then the judgment used to set the criterion may not be appropriate to protect the designated use of the waterbody.

Setting appropriate water quality standards is difficult because it requires a tradeoff between two fundamental objectives: maximizing environmental protection and minimizing the costs of TMDL research and implementation. Science can provide information about which water quality variables could be considered as candidates for nutrient criteria, but the threshold level for the water quality goal can only be set by making explicit tradeoffs between the objectives. At times these tradeoffs are made implicitly, which causes an opaque representation of the difference between the science and value judgments. In contrast, the science and the value judgment can be transparently differentiated, allowing the standard to be properly critiqued for either the scientific basis for the water quality variable or the value judgment dictating the threshold level.

Section 303(c) of the Clean Water Act requires each state to develop water quality standards that are protective of the designated uses of their waters. Therefore, good criteria should be easily measurable and good predictors of the attainment of designated use. The method to accomplish this task, however, is left to the best judgment of the state. States that have adopted nutrient criteria have used a variety of methods. The two most common methods of setting criteria are: (1) the ecoregion reference waterbody strategy, and (2) the use of scientific literature and expert panels.

The U.S. EPA's guidance for setting nutrient criteria is known as the ecoregion reference waterbody strategy (U.S. EPA, 2000a,b, 2001). Using this strategy, a criterion for nutrients is set for a particular ecoregion that is designed to be protective of all its lakes and reservoirs. This criterion is not linked to any of the other variables; a single distribution is created and evaluated for each eutrophication-related water quality variable.

This method is not without flaws. The two greatest flaws are that: 1) it fails to use a statistical analysis to develop predictive criteria, and 2) it fails to distinguish between the science and judgment (i.e., the ecoregion approach, where a particular cutoff level is suggested, seems scientifically driven; however, it is really deciding an arbitrary level that has no grounding in the either science or risk-based values.) (U.S. EPA, 2000b). Thus, the use of reference watersheds is informative for setting criteria for polluted waterbodies only when there are similar pristine, reference watersheds. Most watersheds, however, have multiple land uses, designated uses, and human influences, which makes the choice of a reference waterbody difficult to impossible. Additionally, the use of a 25th percentile distribution point for all waterbodies or 75th percentile distribution point for reference waterbodies is the decision-making judgment, which is often misinterpreted as a strict scientific cutoff point. Therefore, the ecoregion reference waterbody approach is a method to set criteria and threshold levels that uses both scientific knowledge and expert judgment (U.S. EPA, 2000b).

Another common method to set nutrient criteria is to establish nutrient criteria using a combination of eutrophication literature and the selection of the appropriate criteria level by a panel of experts. Such a method was originally used in North Carolina to set nutrient criteria [statewide criterion: 40 μg/L chlorophyll *a* for warm water and 15 μg/L chlorophyll *a* for cold water; see North Carolina Administrative Code, North Carolina Department of Water Quality (NCDWQ), 2003a,b]. This method of setting criterion is similar to the method used by other states, which is a mixture of science-based assessment and expert judgment. One of its flaws, similar to the ecoregion method, is the lack of a direct link to the designated use and the official incorporation of expert judgment.

The team believes that there are alternative and arguably better ways to define the criterion level than through the ecoregion reference waterbody strategy and a panel of experts. Appropriate criteria for eutrophication-related designated use are measurable water quality variables that are statistically predictive of designated use attainment. The level of the criteria, on the other hand, should be chosen based on societal values that reflect the realities of society's tradeoffs between the competing objectives of maximizing environmental protection and minimizing cost (Barnett & O'Hagan, 1997). Beyond that, the selection of the level of the criterion should realistically take into consideration natural variability and uncertainty in predicting water quality outcomes, both of which imply that 100% attainment in space–time is not a realistic basis for a standard.

This search project develops an alternative method to set appropriate nutrient criteria. The focus of the research was to explore the relationships between causal variables and response variables on attainment of a designated use (i.e., the effect of nitrogen and phosphorous on all goal of water clarity response and how both of these relate to designated use attainment). By creating models that focus on the water quality goal, the models indicate criteria that are statistically linked to the designated use.

2.2 Methods to Develop Predictive Nutrient Criteria

The team developed a method that used a combination of eutrophication theory from literature reviews, water quality data, expert elicitation, and SEM. Two of the methods—expert elicitation and SEM—will be discussed in more detail in later chapters. The use of the other methods will be apparent after reading the methods sections and the case study examples.

2.2.1 Expert Elicitation

People tend to think of science as objective, although it is exceedingly rare that subjectivity can be entirely avoided. In fact, judgment is typically necessary throughout a scientific study—from the statement of hypotheses, the specification of a model, the design of experiments or monitoring programs, the selection of methods of analysis, to the final inferences and conclusions—all of these tasks generally involve expert judgment intermingled with the objective analysis of data (Press & Tanur, 2001). Therefore, a realistic appraisal of science must acknowledge a role for the expert judgment of scientists.

Scientists routinely make these judgmental assessments throughout their studies in a thoughtful but informal way. For example, water quality modelers may select the specification of a phytoplankton growth model as a multiplicative function of nitrogen and phosphorus Michaelis–Menten factors, even while recognizing that: 1) other nutrients also may be important, and 2) a limiting nutrient functional form is a plausible alternative. The true growth dynamics are exceedingly complex; the selected model is a pragmatic judgmental choice made by scientists experienced in phytoplankton growth kinetics.

With this perspective in mind, consider the approaches for the selection of water quality standards and criteria presented in the previous section. Certainly, the 1970s strategy involving scientific consensus was heavily judgmental. Most important, while drawing on objective statistical analyses, the proposed U.S. EPA national nutrient criteria development strategy still depends on scientific judgment in the selection of reference conditions. Similarly, the proposed procedure incorporates expert judgment as described above. The need to link measured criteria with narrative designated use statements unavoidably requires expert judgment.

If any strategy employed for nutrient standard setting will have a judgmental component, how best does one elicit and incorporate expert judgment? Fortunately, there is a good answer to that question, because there is a vast literature describing methods for judgmental elicitation.

The goal of expert elicitation is to extract subjective judgments from experts in a systematic procedure. This rigorous, transparent process is frequently used in the decision sciences (Morgan & Henrion, 1990; Meyer & Booker, 1991; Keeney & von Winterfeldt, 1991) because it provides a defensible and well-established method for extracting necessary information. This method of judgmental assessment has also been used in the environmental and aquatic sciences (Reckhow, 1988; Borsuk et al., 2002). The improvement resulting from the use of expert elicitation is that it makes these subjective judgments transparent.

The elicitation methods used in this study were developed based on suggestions from the expert elicitation literature (Morgan & Henrion, 1990; Meyer & Booker, 1991; Keeney & von Winterfeldt, 1991; Borsuk, 2004). Because there is not a single "cookbook" procedure to obtain expert judgment, each expert elicitation procedure differs because of the expert, judgments to be assessed, and project goals. There is, however, a set of adaptive guidelines to ensure that the method would provide us with the best dataset possible.

The initial expert elicitation method used a two-phase procedure (Reckhow et al., 2005). The first phase consisted of a set of five questions that were designed to aid the expert in translating the designated use (see Appendix A). In the first phase, the elicitor also asked the experts, after they provided the water quality variables in question 5, to describe the relationship/structure between the variables. This conceptual model served as the *a priori* model for the structural equation model (see Appendix A). In the second phase of the expert elicitation, the elicitor asked the expert, for a given row of data, what is the probability designated use attainment. These responses were used to quantify designated use attainment within the structural equation models. The appendix includes further detail of the method.

Collection of the necessary data with the previous elicitation approach was relatively easy. The method, however, forced the experts into a linear thought process, which may have confined their thinking. The goal of the interviewer, when asking an expert his or her beliefs, is to reveal their existing values (archeology) instead of structuring desired values (architecture) (Payne et al., 1999). Ideally, the experts would have clearly formed opinions on eutrophication-related designated use attainment, and revealing an expert's values would be straightforward. This assumption is rarely true (Bettman et al., 1998); the methods of questioning or information presentation can have a dramatic impact on an individual's reported values (Coupey et al., 1998). Therefore, the interviewer must be very cautious during questioning to ensure that the questions are unbiased and well articulated to minimize any elicitation bias.

2.2.1.1 Revised Expert Elicitation Protocol

Using what was learned from the previous expert elicitations, the protocol was modified for future nutrient criteria expert elicitations. The revised process includes two parts, both of which can be completed during a one-day meeting (see Appendix B). The first part (approximately 2 hours) will include a discussion about eutrophication processes and designated use impairment as well as the use of expert judgments in this project. In the second part (2 hours of the expert's own time), the expert will provide their judgments about designated use attainment. In this part of the study, the researchers will provide data from a region of waterbodies. For each case, the expert will be given multiple measures of eutrophication-related water quality variables. Given these measures, they will be asked to assess the extent of eutrophication and designated use attainment for that particular lake. The elicitor works through the first few cases with the expert, so they can feel comfortable working through the remaining cases on their own. In total, this process will take approximately 6 hours of the expert's time.

After the team received the expert's judgments, it was summarized in a brief report, which was sent to the experts for approval. The expert was able, at that time, to modify or fine-tune their judgments as they saw fit.

2.2.2 Structural Equation Modeling

Often times in environmental science or ecology, researchers do not want to describe observations or report simple trends in data, but would prefer to find a way to describe the causal structure explaining an underlying a phenomenon (Shipley, 2000). Fortunately, there is a statistical method, Structural Equation Modeling (SEM), that can be used to represent such relationships among variables. Though historically the method has not been used frequently in the natural sciences, more recently, SEM has been increasingly used to represent some of the complexity inherent in environmental systems (Grace & Pugesek, 1997, 1998; McCune & Grace, 2002; Pugesek & Grace, 1998; Smith et al., 1998; Weiher et al., 2004).

SEM was used in this project to represent better the known causal relationships among measured water quality variables. This technique was used to gain insight into the relationship between eutrophication processes and nonattainment of the designated use. This provided knowledge about the water quality variable that would offer the most information as a criterion for assessing designated use attainment. An introduction to this quantitative method and procedure is provided in the remainder of Section 2.2.2.

2.2.2.1 Introduction to Structural Equation Modeling

SEM is a multivariate statistical technique that can be used to describe linear relationships between variables. SEM is a more general extension of multiple regression where the causal relationship between the variables can be described with multiple linear equations. The SEM implies a covariance structure between the variables included in the model. Therefore, SEM is an excellent statistical technique to use when there are processes, such as eutrophication, that can be described through causal interactions that can be represented by the covariance between variables.

SEM has four advantages over multiple regression. SEM can

1. represent the effect of indirect relationships,
2. incorporate measurement error,
3. use latent variables, and
4. use variables collected on categorical scales.

First, SEM allows researchers to model and test both the direct and indirect between variables. Second, you can explicitly incorporate measurement error uncertainty (Bollen, 1989). As a result, because water quality variables and latent variables that are not measured with perfect certainty, the uncertainty can be incorporated into the model so that the model estimates are not biased. Third, latent variables, not just indicator variables, can be used in SEM. Latent variables are unobserved theoretical constructs, method effects, or other phenomena (Kline, 1998). For example, researchers may want to include water clarity in their model. Instead of representing water clarity with a single, imperfect measure, such as Secchi depth, a latent variable titled water clarity could be created that is constructed using several measurable water quality variables such as Secchi depth, turbidity, and total suspended solids. Finally, SEM can include variables that are measured on ordinal or dichotomous scales. Thus, categorical data, such as survey data, can be included and modeled using SEM.

SEM is primarily a confirmatory, rather than exploratory, method. By this, it is meant that a modeler will propose and test a hypothesized model structure that reflects the modeler's scientific understanding of the natural processes and processes. The validity of this structure is then tested by comparing the model-implied covariance structure to the data-implied covariance structure. If the covariance structures are consistent with each other, than the SEM is deemed a plausible representation of the underlying causal structure. If the initial model is not the best representation of the system, then the model is modified, through exploratory techniques, to represent better the studied system. The model validity is then tested using the same procedure described earlier.

More specifically, a structural equation model is fit by minimizing the difference, or residual, between the model-implied covariances matrix and the data-implied covariance structure. Commonly used fitting functions include maximum likelihood, unweighted least squares, and generalized least squares. The model is then tested using a number of different methods that assess the overall fit of the model such as χ^2, Tucker-Lewis Index, and root mean square error (RMSE) of approximation (Bollen, 1989). By looking at the correlation between the parameters, one is able to gain a greater understanding of the relationships inherent in the ecosystem and represented by the measured water quality variables. This offers notable benefits when contrasted to what is learned by examining single variable values.

Because water quality data, in general, are noisy, the data may not strictly adhere to SEM multivariate normality assumption. Therefore, methods that are more robust can improve the predictive power in certain analyses. Such estimators will be called robust estimators because they are robust to departures from multivariate normality of the data (Hampel et al., 1986). If normality is assumed when the distribution is not multivariate normal, then the estimators will behave unexpectedly; this is a result of failures of the distribution assumption that lead to possible bias, inconsistencies, and estimator inefficiencies.

2.2.2.2 Structural Equation Modeling Illustrative Example

As mentioned previously, SEM is an extension of multiple regression where the relationship amongst the variables can be explicitly modeled. For illustrative purposes, consider Figure 2-1.

2.3 Predictive Nutrient Criteria Case Studies

The proposed methodological framework was applied for identifying predictive nutrient criteria using three case studies. The Lake Washington example highlights the basic features of the method, while the Neuse River Estuary application shows how the approach to estimate the probability of use attainment associated with various levels of the candidate criteria can be used and provide graphical expressions of the noncompliance risk (Reckhow et al., 2005). The third case study uses judgments from multiple experts; it is also based on 12 lakes and reservoirs from western North Carolina, and demonstrates the general application of this method at wider geographical scales (e.g., ecoregions, political units).

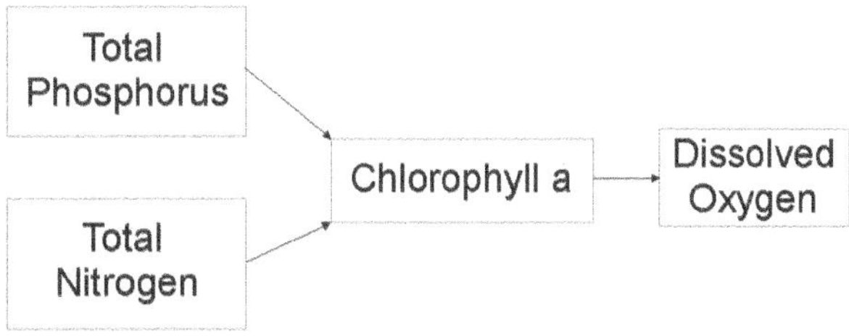

$$Chlorophyll\ a = \alpha + \beta_1\ Total\ Phosphorus + \beta_2\ Total\ Nitrogen + error$$

$$Dissolved\ Oxygen = \alpha + \beta_3\ Chlorophyll\ a + error$$

Figure 2-1. Example Structural Equation Model with the Associated Linear Equations. By extending multiple regression methods and applying SEM, more complex relationships can be represented as a system of equations. In the equations, α represents the intercept and β represents the coefficients.

2.3.1 Lake Washington

Lake Washington is the second largest natural lake in the State of Washington, and is one of the best documented cases of successful restoration by sewage diversion (Edmondson, 1994). Currently, Lake Washington can be characterized as a mesotrophic ecosystem with limnological processes strongly dominated by a recurrent diatom bloom (Edmondson, 1994; Arhonditsis et al., 2003). Lake Washington serves as prime habitat for juvenile salmon and supports both recreational activities and local fisheries (Kerwin, 2001). Washington State submitted a proposed policy for nutrient criteria in 2003, pending approval by the U.S. EPA under Section 303(c) of the Clean Water Act. This nutrient criteria is meant to be protective Lake Washington's designated uses, which are: salmon and trout, primary contact recreation, domestic, water supply, wildlife habitat, commerce and navigation, boating, and aesthetic values (Washington State Department of Ecology, 2003).

Water quality data were obtained from the Major Lakes Monitoring Program in King County, Washington State for 1994–2000. Data on zooplankton abundance and species composition were also obtained from the University of Washington's Department of Biology (Arhonditsis et al., 2004). The expert elicitation was conducted using Dr. Eugene Welch, a Professor Emeritus at the University of Washington, as the expert; he provided data on the attainment of a boating designated use.

Using the water quality and the expert elicitation data, a structural equation model was fit where eutrophication impairment was linked to designated use attainment (Figure 2-2). The final model had relatively good fit and all paths shown are individually significant ($\alpha = 0.05$), except the path between total phosphorus and chlorophyll *a* (Figure 2-2). The model explains 27% of the variation in chlorophyll *a* and 71% of the variation in the probability of designated use attainment (indicated by the numbers in rectangles in Figure 2-2).

There were several interesting observations of the standardized direct effects (the values associated with the arrows in Figure 2-2). The standardized positive path between total phosphorus and chlorophyll *a* (0.20) represents phenomenon that is quite common in the summer epilimnion: an increase in the available phosphorus is associated with the phosphorous being sequestered within the phytoplankton cells. Also, the standardized direct effects of chlorophyll *a* on the probability of designated use attainment were estimated to be -0.59, making chlorophyll *a* directly important in predicting designated use attainment.

To gain a more in-depth understanding of the most predictive, and thus informative, variable, the researchers calculated the direct effects, indirect effects, and total effects (= direct effects + indirect effects) on designated use attainment for both chlorophyll *a* and total phosphorous. For chlorophyll *a*, there are no indirect effects on designated use so the direct and total effects are both -0.59. For total phosphorous, there are direct, indirect, and total effects on designated use attainment; these are -0.43, -0.12 (= 0.20 * -0.59), and -0.55 (= -0.43 + -0.12), respectively. Comparing the total effects of chlorophyll *a* (-0.59) and total phosphorous (-0.55), one can infer that chlorophyll *a* has a somewhat closer association (both direct and total effects) followed by the total phosphorus. Therefore, for this system, chlorophyll *a* would be the most predictive water quality variable.

After developing a plausible model, robust covariance methods were explored to determine if robust techniques would improve the research team's inferences. Minimum covariance determinant and minimum volume ellipsoid methods were used to detect and remove multivariate outliers and then create a covariance matrix that was robust to departures from normality. Comparing the robust methods to the nonrobust methods for Lake Washington, there was no notable difference in the model between the robust (minimum volume ellipsoid: χ^2 *p*-value = 0.110; minimum covariance determinant: χ^2 *p*-value = 0.125) and nonrobust methods (χ^2 *p*-value = 0.097) Consequently, it was decided that it was appropriate to use nonrobust covariances for the SEM, because the use of robust methods did not improve the model development.

2.3.2 Neuse River Estuary

North Carolina's Neuse River Estuary has a long history of excessive algal blooms, bottom-water hypoxia, and fishkills. These problems led the Neuse River to be characterized as

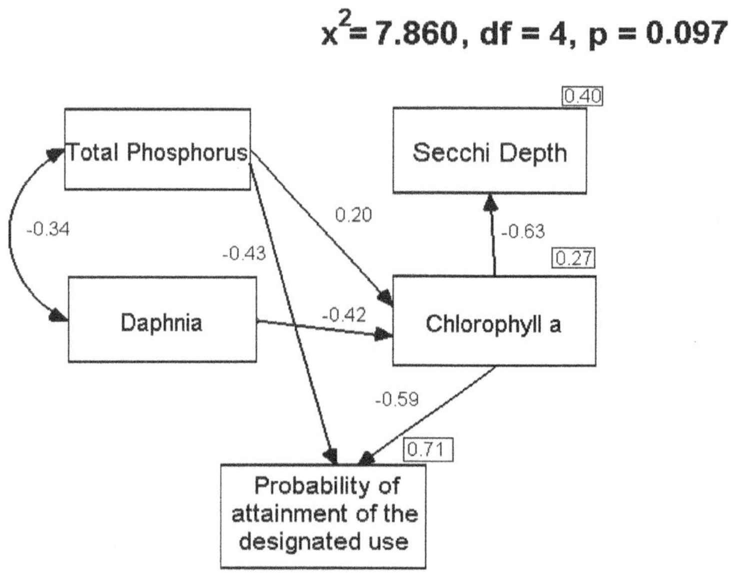

Figure 2-2. Structural Equation Model for Lake Washington. The numbers located near the arrows correspond to the standardized path coefficients and the numbers in rectangles correspond to the R^2 values. x^2, *df,* and *p* correspond to the χ^2 test values, the degrees of freedom, and the probability level for rejecting the null hypothesis, respectively. For purposes of this research, the central message from the figure and its accompanying statistics is that chlorophyll *a* is the most predictive nutrient criterion variable.

one of the 20 most threatened rivers in the United States in 1997 (American Rivers Foundation, 1997). The Neuse has also been listed as an impaired waterbody on the federal 303(d) list because, in certain segments, more than 10% of water quality samples analyzed for chlorophyll *a* exceeded the 40 µg/L criterion. Excessive chlorophyll *a* levels are generally attributed to high point source and non-point source inputs of nitrogen, though developing evidence suggests that phosphorus may sometimes contribute to excessive algal levels (Qian et al., 2000; Paerl et al., 2004). Therefore, in 1997, the NCDWQ developed the Neuse Nutrient Sensitive Waters Management Strategy to reduce total nitrogen loading to the Neuse Estuary by 30% by 2003 (NCDWQ, 1997).

Nutrient criteria are meant to be protective of the designated uses for this estuary. The Neuse River Estuary's designated uses are primary recreation, aquatic life propagation, and maintenance of biological integrity, wildlife, secondary recreation, and any other usage except shellfishing for market purposes (NCDWQ, 2003a,b).

Water quality data were obtained from NCDWQ. Expert elicitation data were collected on the probability of designated use attainment. Dr. Charles Peterson, Professor at the University of North Carolina, Chapel Hill (Institute of Marine Sciences), was the expert.

The final Neuse River Estuary SEM is presented in Figure 2-3, which can be interpreted in a similar way as was indicated for Lake Washington (Section 2.3.1). In this model, DO had a central role, because bottom-water hypoxia often results from increased primary productivity in the euphotic zone that sinks and undergoes bacterial decomposition (Borsuk et al., 2001).

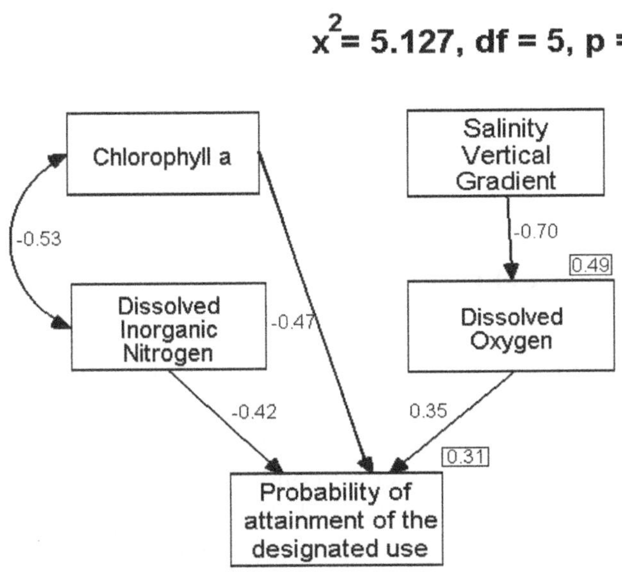

Figure 2-3. Structural Equation Model for the Neuse River Estuary. The numbers located near the arrows correspond to the standardized path coefficients and the numbers in rectangles correspond to the R^2 values. x^2, *df*, and *p* correspond to the χ^2 test values, the degrees of freedom, and the probability level for rejecting the null hypothesis, respectively. For purposes of this research, the central message from the figure and its accompanying statistics is that dissolved oxygen (DO) and chlorophyll *a* are the most predictive nutrient criteria variables.

The model was applied to estimate the probability of use attainment associated with various levels of the candidate criteria. Extending the previous analysis, it was predicted that the probability of designated use surface area using the summer range of DO and chlorophyll *a* levels in the Neuse River Estuary (Figure 2-4*a*); these values were conditional on the concurrent dissolved inorganic nitrogen (DIN) concentrations. In addition, Figure 2-4*a* displays the specific levels of two candidate criteria (DO = 5 mg/L and chlorophyll *a* = 20 µg/L) and assesses their expected probabilities of use attainment, conditional on the values of the other criterion included in the model.

To illustrate the potential use of the probability of attainment analysis, the researchers simplify the graphical representation (Figure 2-4*b*). This graph characterizes the relationship between probability of designated use attainment and the chlorophyll *a* concentration (conditional on DO = 5 mg/L). The probability of designated use expresses the uncertainty in the relationship between the attainment of designated use and the criterion level, which reflects the uncertainty in the elicited expert judgment plus error and variability in the water quality data used to fit the model. The probability of attainment is a realistic quantitative assessment of the research team's ability to assess attainment of the designated use, and is of particular importance to the decision on setting the water quality criterion level.

This example is meant to demonstrate the method, and is not intended as the final analysis leading to a recommended criterion and level. However, one can consider the usefulness of a graph, such as that in Figure 2-4*b*, to examine the choice of chlorophyll *a* level based on risk of nonattainment. For example, the current chlorophyll *a* criterion in the Neuse of 40 µg/L yields only a 60% probability of attainment (Figure 2-4*b*). Similarly, if a decision maker wished to achieve a high likelihood of designated use (>85%) attainment, the chlorophyll *a* criterion would have to be less than 10 µg/L. Of course, absent from this assessment, but critical to standard setting, is a mechanism to allow a decision maker to make direct tradeoffs between the two fundamental objectives of maximizing environmental protection and minimizing TMDL costs and implementation. Nevertheless, this method and representation provides a decision maker with useful information that explicitly represents the tradeoffs in criterion selection that were previously considered implicitly.

2.3.3 Western North Carolina Lakes

Western North Carolina is defined as the area between the political border of North Carolina and the ridge of the Blue Ridge Mountains. In this study, 12 western North Carolina reservoirs were chosen with similar bedrock, soil, and geologic features, where the intersystem variability was mainly driven by differences in local climatic conditions, morphology, and land uses. Compiling the data provided a larger dataset for conducting expert elicitations on the identification of the most predictive criteria and the probability of designated use attainment. [It should be noted that the objective representation of the intrasystem variability was obtained by selecting lakes with more than 15 samplings.] The designated use that was considered for this region of lakes was the propagation and maintenance of a healthy well-balanced population of trout.

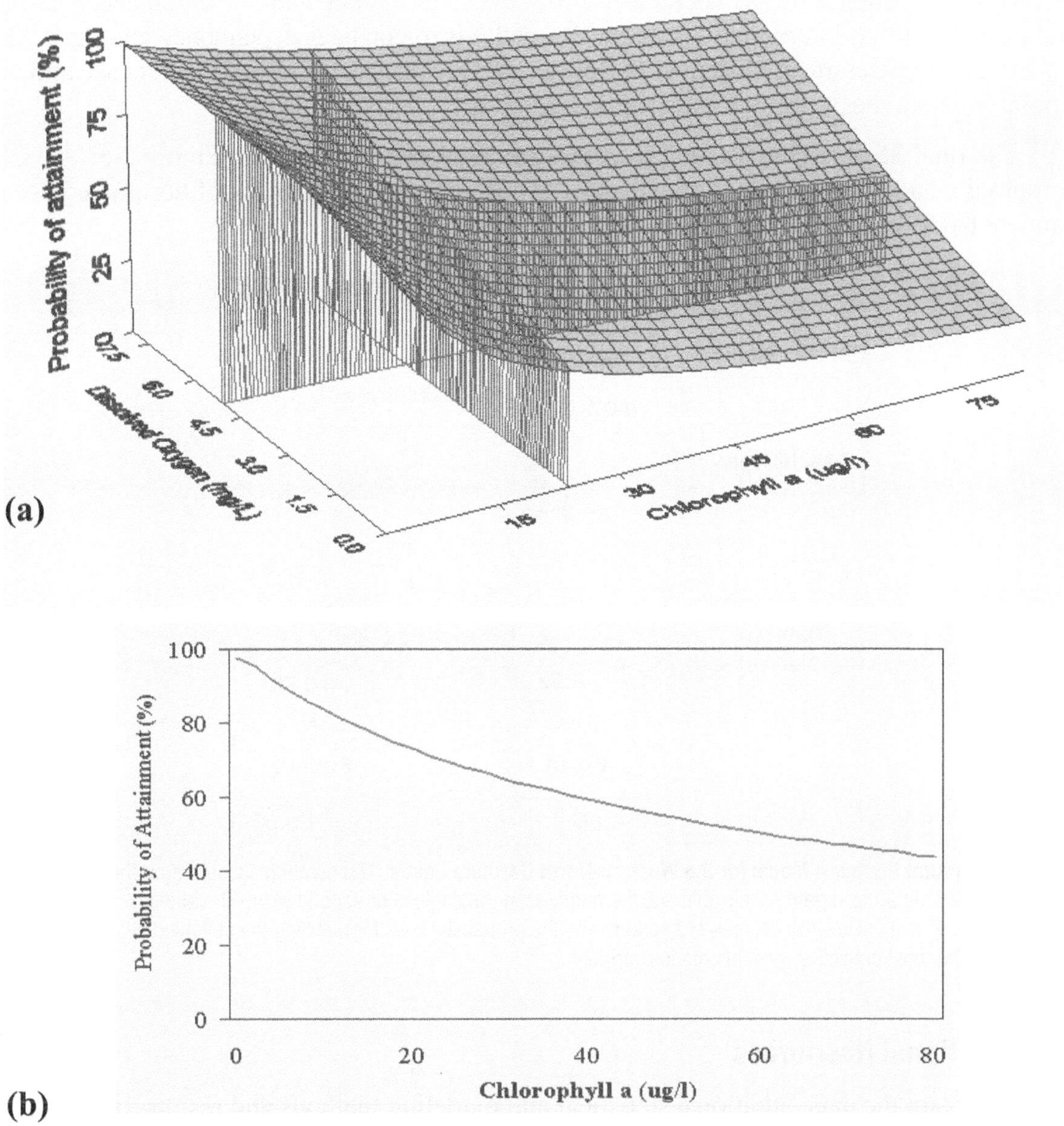

Figure 2-4. Estimation of the Probability of Designated Use Attainment for the Neuse River Estuary. The expert's response of the probability of designated use attainment was logit transformed for this analysis. (*a*) The entire summer range of DO and chlorophyll *a* levels in the Neuse River Estuary. The two shaded/lined surfaces correspond to specific levels of the two candidate criteria: DO = 5 mg/L and chlorophyll *a* = 20 µg/L. (*b*) A two-dimensional presentation of the probability of designated use attainment and chlorophyll *a*, given a constant DO of 5 mg/L.

The data were obtained from NCDWQ from 1981–2003. Data on the probability of designated use attainment were collected from two experts from the North Carolina Department of Natural Resources, NCDWQ, Intensive Survey Unit, Debra Owen and Dianne Reid. Two experts were selected to capture a greater range of designated use interpretations and probability distributions of designated use attainment. Both experts provided their assessment of the probability of designated use attainment for the same 100 data rows. The elicited values from the two experts formed the basis for the respective measurement model (i.e., the latent variable "Designated Use" and the two indicator variables "Expert 1" and "Expert 2").

However, both of these experts were from the same agency and one could argue that they would provide similar interpretations; there are methods to consider dependence amongst experts when combining assessments (Clemen & Reilly, 2001). This case study was a test case and, at this point, it is not meant for nutrient criteria selection.

The final SEM for the western North Carolina lakes is presented in Figure 2-5. Chlorophyll *a* had a large amount of explanatory power for the probability of designated use attainment (-0.98).

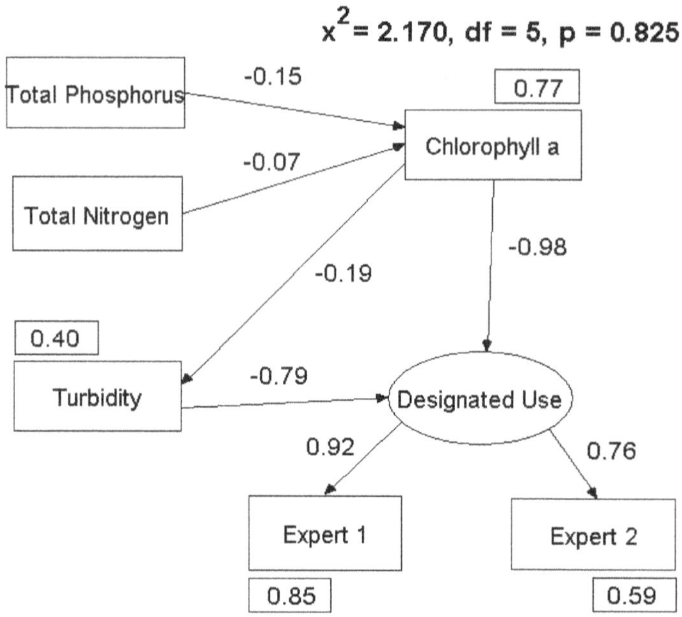

Figure 2-5. Structural Equation Model for the Western North Carolina Lakes. The numbers located near the arrows correspond to the standardized path coefficients and the numbers in rectangles correspond to the R^2 values. x^2, *df*, and *p* correspond to the χ^2 test values, the degrees of freedom, and the probability level for rejecting the null hypothesis, respectively. Chlorophyll *a* is the most predictive nutrient criterion variable.

2.4 Additional Resources

To obtain the data, additional statistical and modeling methods and results, or elaborated descriptions of the methods or case studies described in this section, consult the following references or contact the following individuals:

1. Reckhow, K.H.; Arhonditsis, G.B.; Kenney, M.A.; Hauser, L.; Tribo, J.; Wu, C.; Elcock, K. J.; Steinberg, L.J.; Stow, C.A.; McBride S.J. A Predictive Approach to Nutrient Criteria. *Environ. Sci. Technol.* **2005** *39*, 2913–2919.

2. Kenney, M.A. A Predictive Approach to Selecting Eutrophication-Related Criteria and Threshold Levels for Lakes and Reservoirs in Western North Carolina; Duke University, Durham, N.C., 2005. (Copy available from m.kenney@duke.edu.)

CHAPTER 3.0

ECOLOGICAL PATTERNS

3.1 Overview of Ecological Patterns in Aquatic Systems

Eutrophication research has provided overwhelming evidence that nutrient enrichment should not be analyzed as an isolated causative factor. Nutrient enrichment signals are largely determined by the operational characteristics of individual ecosystems and can further be amplified or dampened by complicated interactions with several other anthropogenic or natural perturbations (Cloern, 2001; Pace, 2001). For example, physiographical properties (bathymetry, basin geography), circulation patterns (tidal and wind mixing, river flows, density stratification), physicochemical attributes of the water column (light availability, salinity), and "top-down" processes (herbivory, benthic–pelagic coupling) change coastal ecosystem dynamics. Thus, these patterns can explain the large differences in ecosystem responses to similar nutrient loadings. Consequently, the major future challenges for the development of effective eutrophication management schemes are: 1) understanding individual and synergistic effects of the various components of the ecosystem filtering function, and 2) ecosystem interactions, given multiple external stressors.

This section assesses the relative role of nutrient concentrations and other biological and nonbiological processes in regulating levels of water quality variables related to designated uses. This objective was accomplished using structural equation models that were applied to a wide range of waterbodies with different trophic states and physical environments. This method of modeling allows the explicit recognition that significant variability exists in terms of the intersystem and intrasystem responses to nutrient enrichment. Two case studies are provided: epilimnetic phytoplankton dynamics during the summer stratification period and phytoplankton patterns in a freshwater–marine continuum, to illustrate how SEM can be used to assess the relative role of important aquatic ecological processes.

3.2 Methods of Structural Equation Modeling

The research team developed a method that used a combination of eutrophication theory from literature reviews, water quality data, and SEM. The use of the method, particularly Bayesian SEM, will be described in more detail. The use of the other methods will be apparent after reading the methods sections and the case studies.

Given the interesting features of SEM, it is quite surprising that SEM applications in ecology and environmental sciences are still limited (e.g., Smith, 1995; Shipley, 1997, 2000; Grace & Pugesek, 1998) and even less common in aquatic ecosystems (Malaeb et al., 2000; Stow & Borsuk, 2003). Thus, in this project, the researchers examine SEM's efficiency to explore ecological patterns. The study shows that this method is well suited for ecological datasets, which are often correlated and error-contaminated, because they provide insight into the relationships of the physical, chemical, and biological variables.

The details of classical SEM were provided in Section 2.2.2.

3.2.1 Bayesian Structural Equation Modeling

For the ecological modeling, a Bayesian approach to SEM was adopted. Bayesian analysis is a statistical method that incorporates prior information with data to create probabilistic assessments. In this research project, a Bayesian hierarchical method was used to make the model identification restrictions stochastic, provide information on multimodality, and relax SEM's asymptotic theory. Because of the flexibility of the Bayesian approach, it is particularly useful in detangling complex ecological patterns.

The Bayesian method has several advantages over the classical methods. First, a Bayesian SEM can incorporate prior knowledge about the parameters and more effectively treat unidentified models (Scheines et al., 1999; Congdon, 2003). Second, the modeling process does not rely on asymptotic theory, a feature that is particularly important when the sample size is small and the classical estimation methods are not robust (Congdon, 2003). In a Bayesian SEM, Markov Chain Monte Carlo samples are taken from the posterior distribution, and consequently, the procedure works for all sample sizes and accommodates for non-normality. Third, the assumptions used to determine the latent variable metrics can be treated stochastically, providing additional insight into the ecological structures (Arhonditsis et al., 2005). Finally, a Bayes factor can be used to evaluate the compatibility of alternative phytoplankton community preconceptualizations with observed ecological patterns (Kass & Raftery, 1995).

3.2.2 Bayesian Structural Equation Modeling References

For further methodological details, consult the following references:

1. Scheines, R.; Hoijtink, H.; Boomsma, A. Bayesian Estimation and Testing of Structural Equation Models. *Psychometrika* **1999**, *64*, 37–52.
2. Congdon, P. *Applied Bayesian Modelling*. Wiley Series in Probability and Statistics, Wiley: West Sussex, England, 2003.
3. Arhonditsis, G.B.; Stow, C.A.; Steinberg, L.J.; Kenney, M.A.; Lathrop, R.C.; McBride, S.J.; Reckhow. K.H. Exploring Ecological Patterns with Structural Equation Modelling and Bayesian analysis. *Ecol. Model.* **2006** *192*, 385–409.
4. Arhonditsis, G.B.; Paerl, H.W.; Valdes, L.M.; Stow, C.A.; Steinberg, L.J.; Reckhow, K.H. Application of Bayesian Structural Equation Modeling for Exploring the Neuse River Estuary (NC, USA) Phytoplankton Dynamics. *Limnol. Oceanogr.,* submitted manuscript.
5. Arhonditsis, G. B.; Stow, C.A.; Paerl, H.W.; Valdes, L.M.; Steinberg, L.J.; Reckhow, K.H. Delineation of the Role of Nutrient Dynamics and Hydrologic Forcing on the Neuse River Estuary (NC, USA) Spatiotemporal Phytoplankton Patterns. *Ecosystems,* submitted manuscript.

3.3 Ecological Patterns Case Studies

In this section the researchers examine the role of nutrients in phytoplankton dynamics in two very different types of situations: 1) in two types of lake systems during periods of summer stratification, and 2) along a freshwater–marine continuum in an estuarine system.

3.3.1 Nutrient Epilimnetic Phytoplankton Dynamics in Lakes

The researchers developed a model to explain epilimnetic phytoplankton dynamics. The model was tested in Lake Washington and Lake Mendota. These lakes were chosen because they have different trophic status; Lake Washington is mesotrophic and Lake Mendota is eutrophic. Also, both of these lakes have a large amount of data, which permitted a deeper level of modeling. These two lakes were used to explore whether the relative role of nutrients in regulating aspects of water quality might differ because of eutrophication status; this phenomenon directly relates to a lake's attainment of designated use.

3.3.1.1 Conceptual Model

A general conceptual model was developed for epilimnetic phytoplankton dynamics (Figure 3-1*a*). Positive results of the conceptual model, when applied to multiple systems, lend credence to the conceptualization of phytoplankton dynamics in general. The SEM that specifically considers the regulatory role of physical, chemical, and biological factors was developed on the response in lake phytoplankton dynamics and water clarity during the summer stratified period (Figure 3-1*b*). The physical factor considered was the change in epilimnion depth. The chemical factor was a latent variable called nutrients, which was indicated by soluble reactive phosphorus (SRP) and DIN. The biological factor was a latent variable entitled herbivory, which was indicated by *Daphnia* and zooplankton. Phytoplankton response, a latent variable called phytoplankton community, was indicated by chlorophyll *a* and either cyanobacteria (Lake Washington) or biovolume (Lake Mendota). Finally, the water clarity response was the Secchi depth measurements.

The application and evaluation of the model was strictly confirmatory; the model was constructed in advance to accommodate current knowledge of the epilimnetic phytoplankton dynamics. No changes were made during the application, and the prespecified model's fit was assessed against the two datasets.

3.3.1.2 Application of the Model to Lake Washington

The conceptual model was applied to Lake Washington to demonstrate whether it is an appropriate general representation of lake phytoplankton community dynamics. Understanding these dynamics is essential to enlighten one's understanding of nutrient criteria development.

The application of the model to Lake Washington, showed fair fit (Table 3-1) when comparing various model fit indices (x^2 test statistic = 27.795, 17 degrees of freedom, *p*-value of 0.047; incremental fit index = 0.858; comparative fit index = 0.834; Hoelter's critical N test = 68, model acceptance at 0.01 significance level) (Bollen, 1989; Bentler, 1990; Hoelter, 1983). These results demonstrate that the conceptual model is a reasonable representation of phytoplankton dynamics. The results of the model, as applied to Lake Washington, are discussed in more detail below.

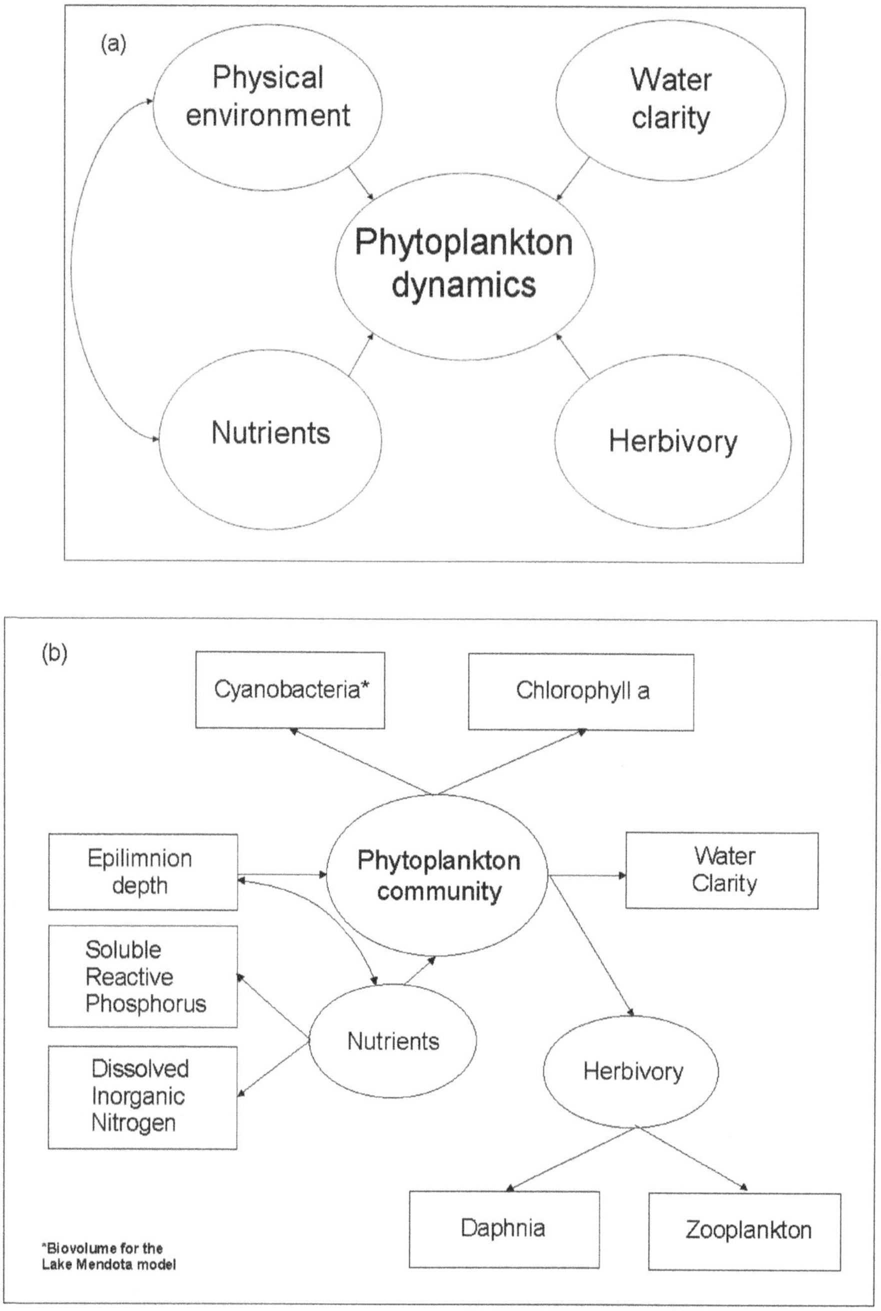

Figure 3-1. Conceptual Models for Epilimnetic Phytoplankton Dynamics. *(a)* The hypothesized conceptualization and (*b*) the actual structural equation model. The use of a rectangular box for the epilimnion depth and the water clarity implies that these variables were considered as directly observable with no measurement error.

The results of the model are presented in Table 3-1. Generally, the hypothesized conceptualization of the epilimnetic phytoplankton dynamics provided satisfactory results that are consistent with existing Lake Washington literature. There were, however, some results that warrant further explanation. The observed variables DIN and SRP were moderately correlated with the latent variable nutrients ($r \approx 0.46$). Chlorophyll *a* was strongly correlated with phytoplankton ($r \approx 0.75$), while cyanobacteria abundance showed a relatively weak negative correlation ($r \approx -0.27$). Both *Daphnia* ($r \approx 0.72$) and total zooplankton ($r \approx 0.89$) length-weighted biomass were strongly correlated with the latent variable herbivory. All the paths between the latent variables of the initial hypothesized structural model were significant with the only exception being the direct path from nutrients to phytoplankton ($p = 0.427$). In addition, the covariance between epilimnion depth and nutrients was positive but nonsignificant ($p = 0.251$). The standardized direct effect of the epilimnion depth on phytoplankton was -0.50. The standardized direct paths from phytoplankton to herbivory (0.55) and water clarity (-0.57) were nearly equal in magnitude, but opposite in sign.

Table 3-1. Epilimnetic Phytoplankton Dynamics—Application to Lake Washington.

Factor Loadings	*Estimate*	R^2
Epilimnion depth	-0.50	
SRP	0.44*	0.20
DIN	0.46*	0.21
Nutrients	0.29	
Daphnia	0.72*	0.51
Zooplankton	0.89*	0.79
Herbivory	0.55*	0.31
Water clarity	-0.57*	0.32
Cyanobacteria	-0.27*	0.07
Chlorophyll *a*	0.75*	0.56
Phytoplankton community		0.24
Covariance		
Epilimnion depth ←→ Nutrients	0.33	

* *p*-value < 0.05, two-tailed

There was no significant link between the inorganic nutrient stock and phytoplankton variability in the Lake Washington summer epilimnion. In addition, the epilimnion lacked significant replenishments from the hypolimnion, which was indicated by a nonsignificant covariance between nutrients and epilimnion depth. This result is likely because of the low nutrient levels below the thermocline (DIN ≤ 300 µg/L and SRP ≤ 15 µg/L; see Lehman, 1988).

There were two regulatory factors for the phytoplankton community structure: mixing processes and grazing pressure imposed by the zooplankton community. First, the negative path from epilimnion depth to phytoplankton is plausible because of dilution effects of epilimnetic erosion and deepening. Second, the positive relationship with herbivory requires further explanation. During the summer stratified period, there is a codependence between the phytoplankton and zooplankton community. In the epilimnion of Lake Washington, zooplankton excretion represents a significant portion of the mixed layer phosphorus supply (60–90%) (Arhonditsis et al., 2004). Thus, zooplankton nutrient recycling fuels phytoplankton growth, which in turn has a positive feedback and sustains herbivore biomass.

It is interesting to note that when the team compared the standardized total effects of nutrients for cyanobacteria ($-0.272 * 0.292 = -0.079$) and chlorophyll *a* ($0.751 * 0.292 = 0.219$), the total effects were different signs. A possible explanation for this difference is a pattern where small, pulsed nutrient inputs from either thermocline migration or external sources selectively subsidize the taxonomic groups of the phytoplankton community with affinity and velocity competitive advantages (e.g., diatoms, chlorophytes) over cyanobacteria (Sommer, 1989).

The model, however, was far from perfect causal representation. It explained a relatively low proportion of the observed variability, and several fit tests were in the area between model acceptance and model rejection. There is one aspect of the model that warrants particular reconsideration and enhancement: soluble reactive phosphorous. In the model, the soluble form of phosphorus was included as the only limiting indicator for latent nutrient variable. In a strongly phosphorus-limited epilimnetic environment, primary producers rapidly uptake soluble reactive phosphorous from the hypolimnion or from zooplankton recycling. As a result, the use of soluble reactive phosphorous data provides limited sensitivity in describing the phosphorus–phytoplankton relationship. The incorporation of an additional indicator variable, which would additionally account for the particulate or organic phosphorus fraction, would likely improve the model.

3.3.1.3 Application of the Model to Lake Mendota

The conceptual model was also applied to Lake Mendota to demonstrate whether it is an appropriate general representation of lake phytoplankton community dynamics. Understanding these dynamics is essential to enlighten one's understanding of nutrient criteria development.

The application of the model to Lake Mendota showed good fit (Table 3-2) when compared to various model fit indices (x^2 test statistic = 22.473, 19 degrees of freedom, p value of 0.261; incremental fit index = 0.989; comparative fit index = 0.987; Hoelter's critical N test = 64, model acceptance at 0.05 significance level) (Bollen, 1989; Bentler, 1990; Hoelter, 1983). These results demonstrate that the conceptual model is a good representation of phytoplankton dynamics; the results of the model, as applied to Lake Mendota, are discussed in more detail below.

The results of the model are presented in Table 3-2. The initial hypothesized structural model was confirmed and all the paths between the latent variables were significant. The only exception was the direct path from epilimnion depth to phytoplankton ($p = 0.436$). The standardized direct effect of nutrients on phytoplankton was -0.841. The standardized direct paths from phytoplankton to herbivory (-0.659) and water clarity (-0.918) were negative and highly significant.

Table 3-2. Epilimnetic Phytoplankton Dynamics—Application to Lake Mendota.

Factor Loadings	*Estimate*	R^2
Epilimnion depth	-0.07	
SRP	0.84*	0.71
DIN	0.99*	0.98
Nutrients	-0.84*	
Daphnia	0.96*	0.93
Zooplankton	0.91*	0.83
Herbivory	-0.66*	0.43
Water clarity	-0.92*	0.84
Biovolume	0.82*	0.67
Chlorophyll *a*	0.89*	0.79
Phytoplankton community		0.76
Covariance		
Epilimnion depth ←→ Nutrients	0.42	

* p-value < 0.05, two-tailed

In contrast with Lake Washington, the Lake Mendota model accounts for a large amount of the observed variability in phytoplankton (76%), herbivory (43%), and water clarity (84%). The latent variable, nutrients, was well correlated with DIN ($r \approx 0.99$) and SRP ($r \approx 0.84$). Chlorophyll *a* and total algal biovolume were also highly correlated with phytoplankton and the respective coefficients were 0.89 and 0.81. In addition, both *Daphnia* ($r \approx 0.96$) and total zooplankton ($r \approx 0.91$) length-weighted biomass were strongly correlated with the latent variable herbivory.

The significant correlation between epilimnion depth and nutrients is consistent with previous Lake Mendota studies that emphasize the contribution of internal loading to the epilimnetic nutrient stock. Soranno (1997) found that major entrainment events occur and result in nutrient fluxes that are significantly higher than fluxes from external nutrient sources. These nutrient pulses stimulate phytoplankton (cyanobacteria) blooms (chlorophyll $a \geq 30$ μg /L) and surface scum formation, which result in adverse affects on water clarity and lake aesthetics (Lathrop et al., 1996; Soranno, 1997).

The model results also indicated that herbivory controls summer epilimnetic phytoplankton dynamics. This result is consistent with several past Lake Mendota studies. These studies were based on short- and long-term datasets, variant temporal resolution (from days to seasonal averages), and contemporaneous or lagged measurements showed similar evidence, which underscore the role of zooplankton grazing pressure (especially from the large-bodied *Daphnia* spp.) on phytoplankton dynamics (e.g., composition, abundance) and water clarity (Lathrop et al., 1996, 1999; Soranno, 1997).

3.3.2 Nutrient Epilimnetic Phytoplankton Dynamics in a Freshwater–Marine Continuum

A model was developed to explain phytoplankton dynamics in a freshwater–marine continuum. This was developed to understand better the nutrient and phytoplankton relationships; an improved understanding of the relationships leads to better nutrient criteria development. The model was tested in the Neuse River Estuary, which spans a freshwater–marine continuum.

3.3.2.1 Conceptual Model

A general conceptual model was developed for the Neuse River Estuary (Figure 3-2). The modeling framework delineated the hydrological forcing with regard to nutrient role on phytoplankton growth and compositional alterations along the estuary. This model attempted to evaluate the relative importance of the physical environment (flow, salinity, temperature, and light availability), inorganic nitrogen, and phosphate effects on the spatiotemporal phytoplankton biomass and community composition patterns. The total phytoplankton dynamics were described in the model using a latent variable, titled phytoplankton, which was indicated by chlorophyll *a* and primary productivity.

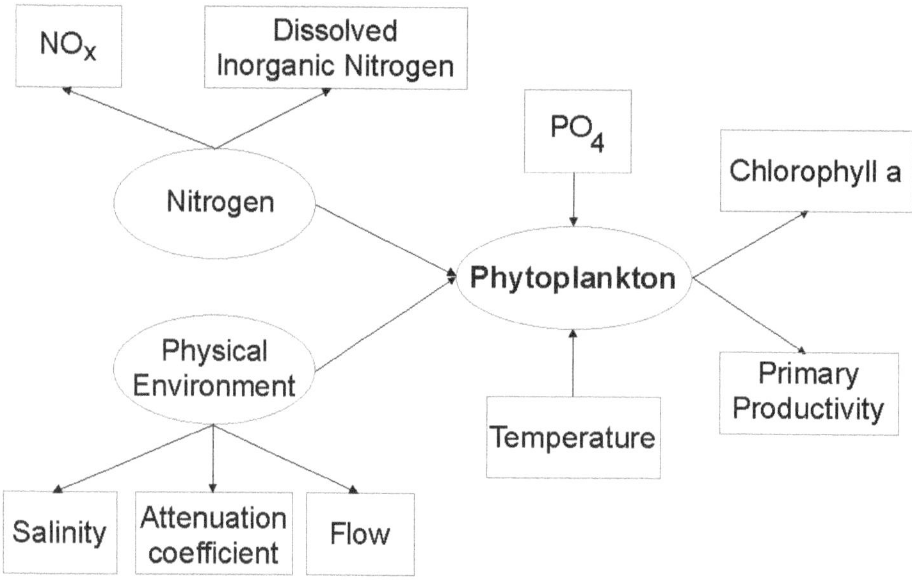

Figure 3-2. Conceptual Model for Phytoplankton Dynamics in a Freshwater–Marine Continuum. The structural equation model used to predict phytoplankton dynamics in the Neuse River Estuary. The use of rectangular boxes for temperature and phosphate implies that these variables were considered as directly observable with no measurement error.

The research team obtained a comprehensive Neuse River Estuary dataset that included data on five major phytoplankton taxa dinoflagellates, diatoms, cyanobacteria, cryptophytes, and chlorophytes. These data allowed the research team's to expand its original conceptualization and examine the relative importance of the various abiotic factors on the phytoplankton community composition. Using diagnostic photopigment data, researchers have previously evaluated as the importance of various phytoplankton indicators of coastal water quality and trophic state (Paerl et al., 2003, 2005).

To capture some of ecosystem complexity, the conceptual model was expanded to consider several phytoplankton taxa groups (Figure 3-3). Four models were constructed to allow several phytoplankton taxa to characterize different phytoplankton functional groups (PFGs). The first model creates PFG A, comprised of diatoms, cryptophytes, and chlorophytes. The first model also assumes that dinoflagellates and cyanobacteria behave individually (Figure 3-3a). The second model creates PFG B and PFG C. PFG B aggregates cryptophytes and chlorophytes; PFG C aggregates dinoflagellates and diatoms. The second model considers cyanobacteria separately (Figure 3-3b). The third model creates two functional groups, PGF D and PFG E. PGF D combines chlorophytes and diatoms; PGF E combines dinoflagellates and cryptophytes. Cyanobacteria were assumed to behave individually (Figure 3-3c). Finally, the fourth model considers two functional groups, PGF A and PFG F, a new functional group that aggregates dinoflagellates and cyanobacteria (Figure 3-3d).

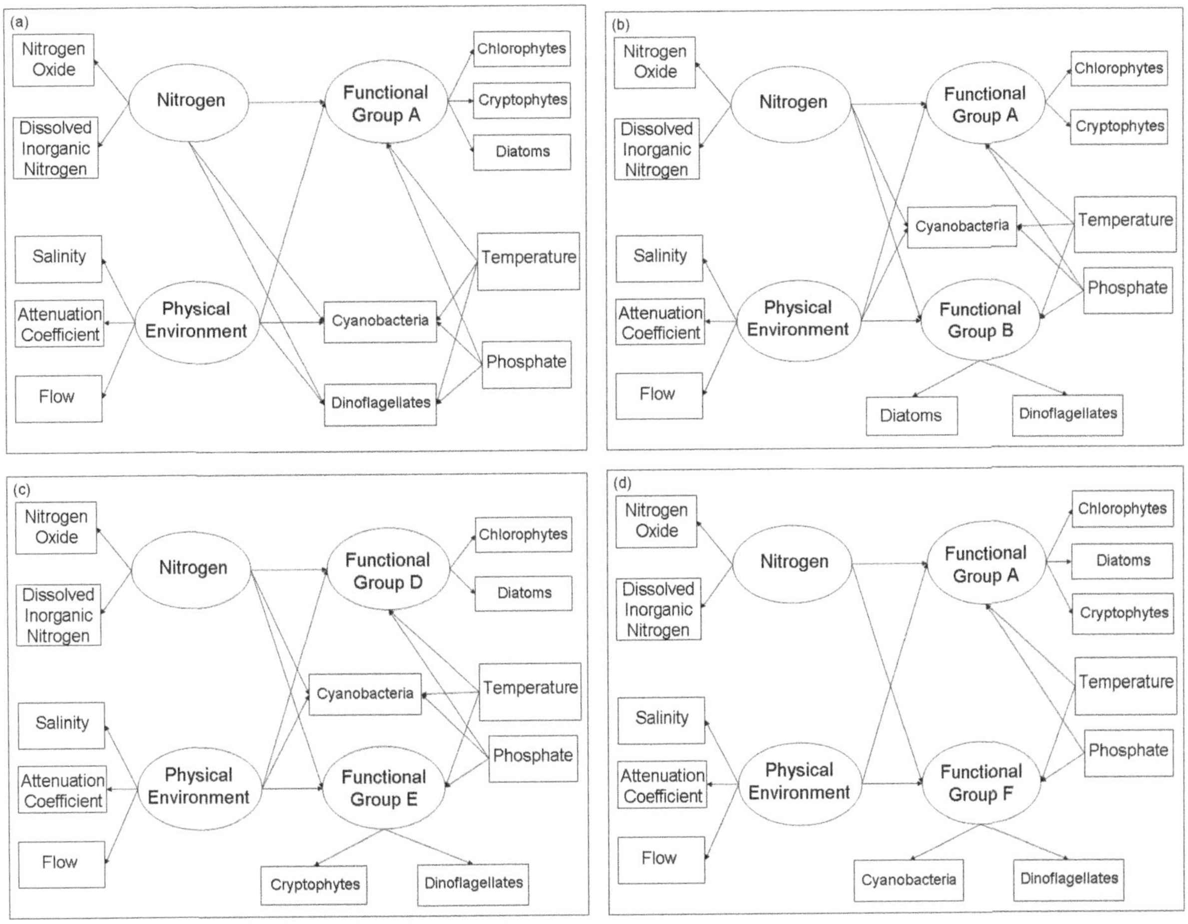

Figure 3-3. The Four Conceptual Models of Phytoplankton Community Dynamics for the Neuse River Estuary.

3.3.2.2 Observed Spatiotemporal Patterns

To account for the spatiotemporal variability within the Neuse River Estuary, four spatial segments were explicitly considered (Figure 3-4). Examining the 1995–2001 data, the researchers discovered different patterns and trends in the different segments. In the upper estuary (segment A), the average chlorophyll *a* levels were lower (≈ 3.29 µg/L) than the lower segments. An additional pattern showed that chlorophytes and cryptophytes were the predominant groups (>30%), followed by diatoms (18%) and cyanobacteria (12%) (Table 3-3). In the mid estuary (segments B and C), there were higher average chlorophyll *a* concentrations (11.74 and 16.71 µg/L, respectively), and diatoms, chlorophytes, and cryptophytes consistently comprised 65–70% of the total phytoplankton community biomass. Finally, moving downstream, there were increasing trends in both cyanobacteria and dinoflagellate concentrations and average percentage contribution; segment D, is dominated by cyanobacteria and dinoflagellate concentrations [average concentration of 2.74 µg/L (≈ 32%)].

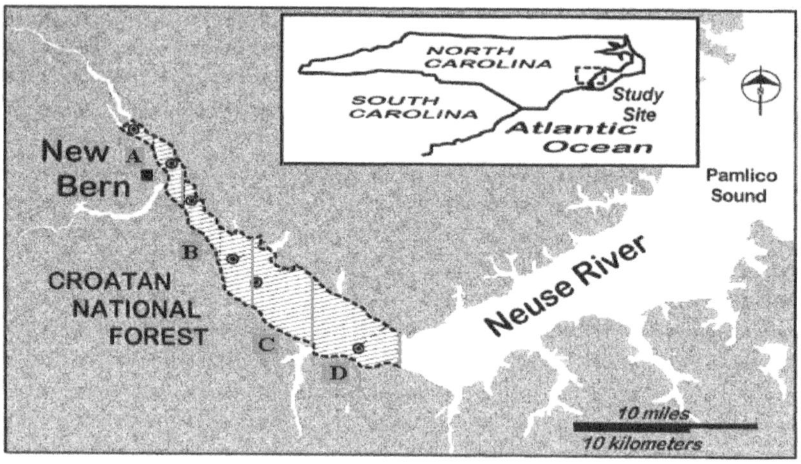

Figure 3-4. Four Spatiotemporal Segments within the Neuse River Estuary. The vertical lines separate the four segments used for this study.

Table 3-3. Average Chlorophyll *a* and Phytoplankton Taxa Values in the Four Segments of the Neuse River Estuary Study Site. Numbers in parentheses indicate the average percentage contribution of the five phytoplankton taxa.

	Section A	Section B	Section C	Section D
Chlorophyll *a*	3.29	11.74	16.71	9.32
Chlorophytes	0.80 (35%)	2.14 (23%)	2.73 (19%)	1.31 (16%)
Cryptophytes	1.03 (32%)	3.47 (32%)	3.69 (23%)	1.66 (18%)
Cyanobacteria	0.35 (12%)	2.52 (18%)	4.10 (26%)	2.74 (32%)
Diatoms	0.86 (18%)	2.52 (20%)	3.49 (21%)	1.95 (20%)
Dinoflagellates	0.25 (3%)	1.39 (7%)	2.36 (11%)	1.76 (14%)

Upstream DIN concentrations showed a fairly steady annual cycle with an approximate median value of 600 μg/L (Figure 3-5; see also Arhonditsis, Stow et al., submitted). The DIN annual median levels were considerably lower in the other three segments (≈ 379, 104, and 46 μg/L for segments B, C, and D, respectively), where the annual minima usually occur during the summer months (July–August). In particular, down-estuary DIN concentrations were almost always consistently lower than 50 μg/L from April to October. In contrast, phosphate concentrations decreased downstream (up-estuary section = 42 μg/L; down-estuary section = 18 μg/L). The annual patterns were characterized by distinct late summer-early fall annual maxima. The DIN: dissolved inorganic phosphorus (DIP) molar ratios were higher in the upstream area (annual median values in sections A and B > 24) and dropped below the Redfield ratio (16; Redfield, 1958) after the middle part of the estuary (annual median values C and D < 10). In addition, the DIN:DIP patterns were fairly similar along the estuary with wide seasonal variation and annual minimum values usually observed by the end of summer (Figure 3-5).

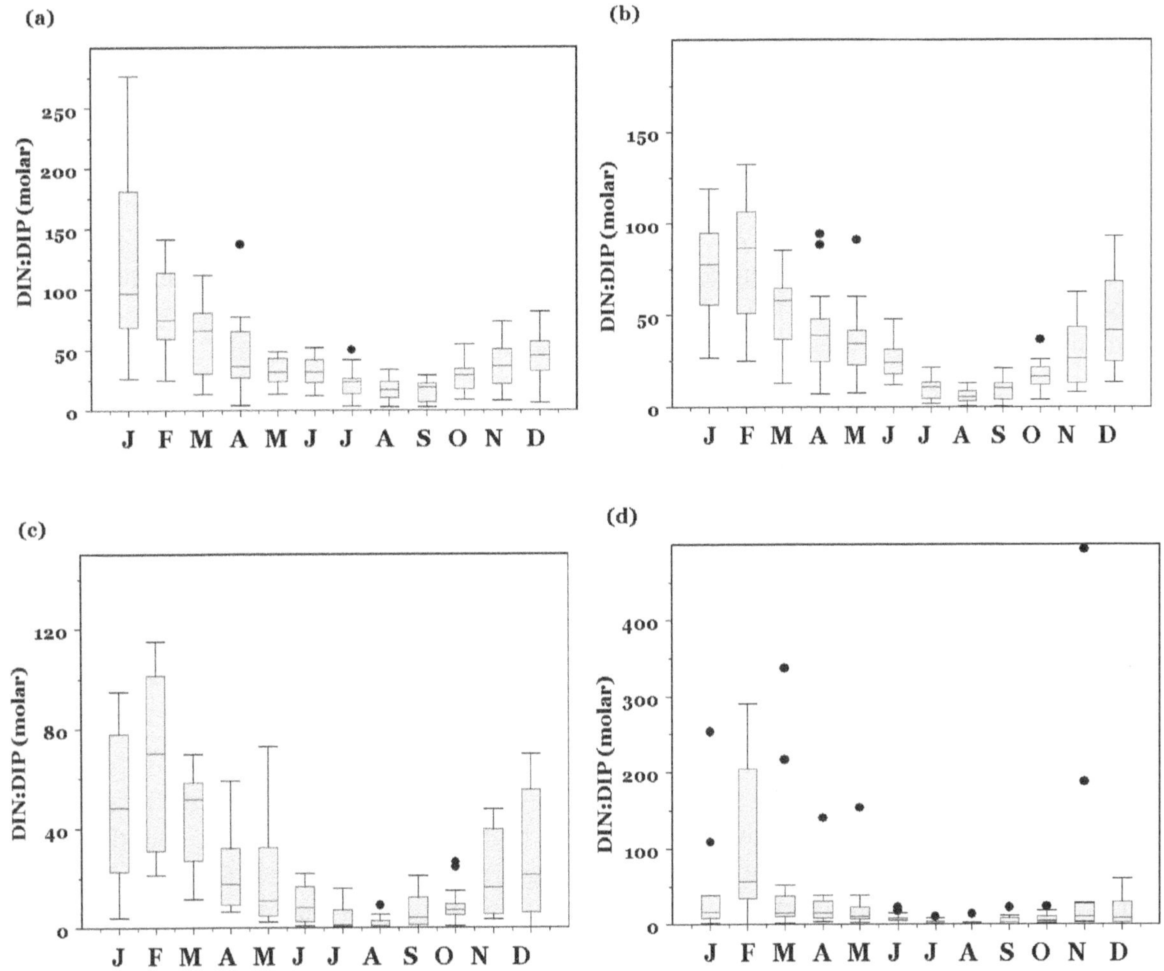

Figure 3-5. Seasonal Variability of the DIN:DIP (Molar) Ratio in the Four Segments of the Neuse River Estuary. The respective median annual values were 31, 24, 10, and 7.

3.3.2.3 Model Results and Discussion

The compatibility of the four different conceptual models was tested with the observed ecological patterns. The model that performed the best, on average, was selected for each spatial section; this was done for optimal phytoplankton community structure aggregation. Thus, the research team's framework determines which site-specific PFGs reflect the regulatory role of abiotic factors and biological interactions. This structure is meant to address some of the weaknesses of the conventional phylogenetically based classification schemes. The model performances of two conceptualizations presented previously are compared: 1) model of phytoplankton dynamics (Figure 3-2), and 2) model of phytoplankton community dynamics with PFGs (Figure 3-3).

The conceptual model (Figure 3-2) was a plausible model with reasonable fit (Table 3-4). Because Bayesian SEM was applied, RMSE was used instead of R^2 values. RMSE is the difference between the predictive value and the actual value; the smaller the RMSE, the better the model performance. In the model (Figure 3-2, Table 3-4), the range of values in the four segments was 2.5–7.6 μg/L for chlorophyll *a* and 4–23 mg C $m^{-3}h^{-1}$ for primary productivity. In addition, the RMSE values for the two forms of inorganic nitrogen varied from 11 to 30 μg/L. Finally, the RMSE values for salinity, attenuation coefficient, and flow were from 1.32–2.65%, 0.38–0.47 m^{-1}, and 38–57 m^3/s, respectively.

The direct standardized path from nitrogen to phytoplankton was weak in the upper part of the estuary (A = -0.02) and became stronger moving to the down-estuary segments (B = -0.08, C= -0.25, and D = -0.26). In contrast, the direct path from phosphorus to phytoplankton was stronger in the upper- and mid-estuary segments (A = -0.17, B = -0.28, and C = -0.16), and weaker down-estuary (D = -0.09). Temperature effects on phytoplankton became weaker moving down-estuary (i.e., from 0.49 to 0.13). Also, the salinity has a higher absolute loading on the physical environment latent variable in the middle- and down-estuary segments C and D.

The physical environment played a significant role on phytoplankton dynamics in the up-estuary segments A and B. In particular, the flow rates have the stronger total standardized effects -0.39 (= -0.43 x 0.91) and -0.64 (= -0.69 x 0.93) among the three indicator variables used for the physical environment measurement model. The path from the physical environment to phytoplankton was weaker in the segment C (-0.13), and the posterior median effects were switched from negative to positive (0.11) in the spatial section D.

The four conceptual models (Figure 3-3) were run for the Neuse River Estuary; the model that considers phytoplankton community comprised of PFG A (diatoms, chlorophytes, and cryptophytes), cyanobacteria, and dinoflagellates (Figure 3-3a) provided satisfactory results (Table 3-5).

The RMSE for the five group-specific chlorophyll *a* values were lower in the upper part of the estuary (<1 μg/L) than in the lower portion. The highest RMSE values were found for cryptophytes and cyanobacteria in the section B (2.2 μg/L), for dinoflagellates (3.2 μg/L), cyanobacteria (2.5 μg/L) in the section C of the estuary, and for dinoflagellates (1.8 μg/L), cyanobacteria (1.9 μg/L) in the down-estuary section D. The RMSE values for the two inorganic nitrogen forms varied from 5 to 38 μg/L; salinity, attenuation coefficient, and flow ranged from 1.42–2.52%, 0.39–0.43 m^{-1}, and 43–63 m^3/s, respectively.

Among the three phytoplankton functional groups of the model (dinoflagellates, cyanobacteria, and PFG A), dinoflagellates had the relatively stronger posterior median path with nitrogen (-0.17) in the up-estuary section A, whereas the respective cyanobacteria and PFG A–nitrogen paths were fairly weak (<-0.09). Coupling dinoflagellates (-0.34) and cyanobacteria (-0.25) with nitrogen was strong in segment B. In addition, the dinoflagellates had the strongest association with nitrogen in the mid- and down-estuary sections C and D (-0.37 and -0.24), where the respective PFG A and cyanobacteria path values were relatively similar and constant (≈ -0.20). The posterior median standardized phosphate–PFG A paths were strongly negative in the upper and middle sections of the estuary (<-0.26), and relatively weaker in the downstream section D (-0.15). Likewise, the up- and mid-estuary cyanobacteria–phosphate paths were negative (<-0.14), whereas their coupling with phosphorus reduced moving downstream. Interestingly, the dinoflagellates seem to have a positive, but relatively weak, relationship with phosphate along the estuary (0.04–0.18).

Table 3-4. Spatiotemporal Structural Equation Model Results for the Neuse River Estuary General Phytoplankton Community Dynamics.

Factor Loadings	Estimate	RMSE
a)		
Nitrogen	-0.02	
Phosphate	-0.17	
Chlorophyll *a*	0.73	2.5 μg/L
Primary productivity	0.96	4 mg C $m^{-3}h^{-1}$
Temperature	0.49	
Physical environment	-0.43	
Nitrogen oxide (NOX)	0.98	27 μg/L
DIN	0.97	28 μg/L
Salinity	-0.50	1.65%
Attenuation coefficient	0.47	0.47 m^{-1}
Flow	0.91	38 m^3/s
b)		
Nitrogen	-0.08	
Phosphate	-0.28	
Chlorophyll *a*	0.86	5.8 μg/L
Primary productivity	0.83	23 mg C $m^{-3}h^{-1}$
Temperature	0.45	
Physical environment	-0.69	
NOX	0.97	30 μg/L
DIN	0.99	15 μg/L
Salinity	-0.71	2.65%
Attenuation coefficient	0.49	0.43 m^{-1}
Flow	0.93	38 m^3/s
c)		
Nitrogen	-0.25	
Phosphate	-0.16	
Chlorophyll *a*	0.71	7.6 μg/L
Primary productivity	0.91	14 mg C $m^{-3}h^{-1}$
Temperature	0.30	
Physical environment	-0.13	
NOX	0.97	25 μg/L
DIN	0.99	11 μg/L
Salinity	-0.92	1.32%
Attenuation coefficient	0.69	0.38 m^{-1}
Flow	0.84	57 m^3/s
d)		
Nitrogen	-.26	
Phosphate	-0.09	
Chlorophyll *a*	0.73	5.1 μg/L
Primary productivity	0.92	11 mg C $m^{-3}h^{-}$
Temperature	0.13	
Physical environment	0.11	
NOX	0.95	27 μg/L
DIN	0.97	13 μg/L
Salinity	-0.92	1.40%
Attenuation coefficient	0.74	0.38 m^{-1}
Flow	0.87	48 m^3/s

Table 3-5. Spatiotemporal Structural Equation Model Results for the Neuse River Estuary Phytoplankton Community Dynamics.

Factor Loadings	*Estimate*	*RMSE*
a)		
Nitrogen→NOX	0.96	35 µg/L
Nitrogen→DIN	0.99	16 µg/L
Nitrogen→PFG A	-0.09	
Nitrogen→Dinoflagellates	-0.17	
Nitrogen→Cyanobacteria	-0.03	
Phosphate→PFG A	-0.26	
Phosphate→Dinoflagellates	-0.09	
Phosphate→Cyanobacteria	-0.015	
Physical environment→PFG A	-0.41	
Physical environment→Dinoflagellates	-0.40	
Physical environment→Cyanobacteria	-0.38	
Physical environment→Salinity	-0.54	1.60%
Physical environment→Attenuation coefficient	0.45	0.43 m^{-1}
Physical environment→Flow	0.84	63 m^3/s
PFG A → Chlorophytes	0.86	0.5 µg/L
PFG A→Diatoms	0.83	0.7 µg/L
PFG A→Cryptophytes	0.73	0.9 µg/L
Temperature→PFG A	0.23	
Temperature→Dinoflagellates	-0.14	
Temperature→Cyanobacteria	0.40	
b)		
Nitrogen→NOX	0.96	38µg/L
Nitrogen→DIN	0.99	5 µg/L
Nitrogen→PFG A	-0.07	
Nitrogen→Dinoflagellates	-0.34	
Nitrogen→Cyanobacteria	-0.25	
Phosphate→PFG A	-0.44	
Phosphate→Dinoflagellates	0.09	
Phosphate→Cyanobacteria	-0.02	
Physical environment→PFG A	-0.42	
Physical environment→Dinoflagellates	-0.58	
Physical environment→Cyanobacteria	-0.51	
Physical environment→Salinity	-0.75	2.52%
Physical environment→Attenuation coefficient	0.52	0.42 m^{-1}
Physical environment→Flow	0.84	58 m^3/s
PFG A → Chlorophytes	0.85	1.2 µg/L
PFG A→Diatoms	0.87	1.5 µg/L
PFG A→Cryptophytes	0.77	2.2 µg/L
Temperature→PFG A	0.45	
Temperature→Dinoflagellates	0.20	
Temperature→Cyanobacteria	0.38	
c)		
Nitrogen→NOX	0.97	32 µg/L
Nitrogen→DIN	0.99	6 µg/L
Nitrogen→PFG A	-0.17	
Nitrogen→Dinoflagellates	-0.37	
Nitrogen→Cyanobacteria	-0.19	
Phosphate→PFG A	-0.34	
Phosphate→Dinoflagellates	0.18	
Phosphate→Cyanobacteria	-0.14	
Physical environment→PFG A	0.017	
Physical environment→Dinoflagellates	-0.20	

Physical environment→Cyanobacteria	-0.22	
Physical environment→Salinity	-0.92	1.42%
Physical environment→Attenuation coefficient	0.69	0.40 m^{-1}
Physical environment→Flow	0.84	61 m^3/s
PFG A→Chlorophytes	0.85	1.2 µg/L
PFG A→Diatoms	0.71	2.1 µg/L
PFG A→Cryptophytes	0.62	2.0 µg/L
Temperature→PFG A	0.11	
Temperature→Dinoflagellates	0.29	
Temperature→Cyanobacteria	0.55	
d)		
Nitrogen→NOX	0.94	31 µg/L
Nitrogen→DIN	0.98	12 µg/L
Nitrogen→PFG A	-0.21	
Nitrogen→Dinoflagellates	-0.24	
Nitrogen→Cyanobacteria	-0.19	
Phosphate→PFG A	-0.15	
Phosphate→Dinoflagellates	0.04	
Phosphate→Cyanobacteria	0.00	
Physical environment→PFG A	0.36	
Physical environment→Dinoflagellates	0.14	
Physical environment→Cyanobacteria	0.01	
Physical environment→Salinity	-0.91	1.55%
Physical environment→Attenuation coefficient	0.74	0.39 m^{-1}
Physical environment→Flow	0.88	43 m^3/s
PFG A→Chlorophytes	0.71	0.7 µg/L
PFG A→Diatoms	0.81	0.9 µg/L
PFG A→Cryptophytes	0.52	1.1 µg/L
Temperature→PFG A	-0.35	
Temperature→Dinoflagellates	-0.09	
Temperature→Cyanobacteria	0.37	

The model results indicated that the physical environment is the predominant factor that regulates phytoplankton dynamics at the upper reaches of the estuary (sections A and B), where the respective posterior median paths for both the aggregate and compositional models were consistently strong (<-0.40). Given the high standardized flow rate loading values on the latent physical environment, the researchers also infer that the phytoplankton dynamics in the relatively narrow upper Neuse River Estuary are primarily driven by the control of flow on the phytoplankton growth-minus-physical advection loss balance (Twomey et al., 2002; Borsuk et al., 2004). As a result, the up-estuary total phytoplankton maxima usually occur during periods of low flow conditions [i.e., late summer–early fall (section A) and late spring–mid summer (section B)], which is also reflected by the positive temperature path values (>0.45). In addition, the chlorophyll *a* concentrations are notably higher as the researchers move downstream (Pinckney et al., 1997).

In summary, a Bayesian SEM framework was used to explore the Neuse River Estuary spatiotemporal phytoplankton aggregate and compositional patterns. Resolution of spatiotemporal nutrient limitation patterns versus hydrological forcing effects on algal growth along a freshwater–marine continuum is essential for guiding the costly implementation of nutrient reduction strategies and optimizing their effectiveness. The model highlighted the role of the physical environment (river flow fluctuations and salinity and light availability) that dominates the up-estuary processes, loosening the coupling between nutrient and phytoplankton.

The switch from an upstream negative to a downstream positive phytoplankton–physical environment relationship suggests that the elevated advective transport from the upper reaches of the estuary can lead to a phytoplankton biomass accumulation in the mid- and down-estuary segments. The positive influence of the physical environment on the phytoplankton community response was more evident on the respective diatom, chlorophyte, and cryptophyte paths, highlighting the opportunistic behavior of these taxa. Diatom, chlorophyte, and cryptophyte have rapid nutrient uptake and growth rates and a high tolerance to low salinity conditions that allows them to dominate the phytoplankton community during high freshwater conditions. Both the model results and the ambient nutrient concentrations indicated: 1) a stronger association between phosphorus and total phytoplankton dynamics in the upstream freshwater locations, 2) both nitrogen and phosphorus played a significant role in the middle section of the estuary, and 3) the nitrogen–phytoplankton paths showed the highest absolute values in the downstream mesopolyhaline zone. This nutrient limitation pattern was seen primarily in the diatom, chlorophyte, and cryptophyte dynamics. Finally, the researchers' analysis demonstrated that as they move to the down-estuary segment, cyanobacteria dominate the phytoplankton community because of a protracted favorable environment (e.g., longer residence times, low DIN concentrations, and relaxation of the phosphorus limitation); this results in temporal pattern shifts on the phytoplankton community structure.

Finally, the results of the study of the ecological patterns in aquatic systems have a tremendous amount of importance for nutrient criteria. The role of nutrients change throughout the freshwater–marine continuum. In freshwater, phosphorous plays the predominant role, in the middle of the estuary both nitrogen and phosphorous are important. Nitrogen has the largest role in the estuary. The research team concluded that it is important to consider the location of the waterbody when setting nutrient criteria.

3.4 Additional Resources

To obtain the data, additional statistical and modeling methods and results, or elaborated descriptions of the methods or case studies described in this section, please consult the following references or contact the following individuals:

1. Arhonditsis, G.B.; Stow, C.A.; Steinberg, L.J.; Kenney, M.A.; Lathrop, R.C.; McBride, S.J.; Reckhow, K.H. Exploring Ecological Patterns with Structural Equation Modelling and Bayesian Analysis. *Ecol. Model.* **2006** *192*, 385–409.
2. Arhonditsis, G.B.; Paerl, H.W.; Valdes, L.M.; Stow, C.A.; Steinberg, L.J.; Reckhow, K.H. Application of Bayesian Structural Equation Modeling for Exploring the Neuse River Estuary (NC, USA) Phytoplankton Dynamics. *Limnol. Oceanogr.,* submitted manuscript.
3. Arhonditsis, G.B.; Stow, C.A.; Paerl, H.W.; Valdes, L.M.; Steinberg, L.J.; Reckhow, K.H. Delineation of the Role of Nutrient Dynamics and Hydrologic Forcing on the Neuse River Estuary (NC, USA) Spatiotemporal Phytoplankton Patterns. *Ecosystems,* submitted manuscript.

CHAPTER 4.0

Conclusions and Suggestions for Future Work

4.1 Research Conclusions

The objective of this project was to develop an alternative and more appropriate method to select nutrient criteria. This research is particularly important because it offers scientific basis for nutrient criteria that is predicatively linked to designated use attainment.

This research proposes an alternative method to set appropriate nutrient criteria that is a mix of science and expert judgment. In this regard, the researchers' method has much in common with portions of the analysis recommended by U.S. EPA. It is analytically thorough and rigorous, involving data analysis, modeling, and expert judgment. This approach, however, avoids the value judgment implied in the reference conditions. This research develops a methodological framework that predicts the probability of designated use attainment as a function of criterion level; it leaves the choice of criterion level to policy makers. Once the procedure was developed, the major challenges in this work were the quantification of the narrative designated use statement and the selection of a modeling strategy that accurately describes the causal ecological structures underlying the system function. To address these research issues, the researchers chose expert elicitation for the former and SEM for the latter. The case studies demonstrated the promise of this method, which is still being refined because of ongoing research.

In directing the U.S. EPA to develop a national strategy for nutrient criteria, the President's 1998 Clean Water Action Plan lays the foundation for addressing the leading cause of TMDLs nationwide. Unfortunately, setting appropriate nutrient criteria is not a straightforward process because there are both complex eutrophication processes that dictate the likelihood of waterbody impairment and the inherent tradeoff between environmental protection and TMDL research and implementation costs. This research was meant to shed light on the former problem. The combination of expert elicitation, user surveys, and SEM that the predictive strategy encompasses has several advantages over the U.S. EPA approach.

4.2 Funded Future Research Extensions

The North Carolina Department of Natural Resources, Division of Water Quality (NCDWQ) has provided Kenneth H. Reckhow, Melissa A. Kenney, and George B. Arhonditsis funding to further explore the method of predictive nutrient criteria and develop a recommendation for the State of North Carolina. More specifically, the focus of the research project is to explore the relationships between the causal variables, such as nitrogen and phosphorus, and the response variables, such as chlorophyll *a* and turbidity, on attainment of a designated use in nonflowing waters in North Carolina. Because each waterbody and associated watershed is unique, this method focuses on the goal and water quality parameters that are similar in a set of waterbodies. This study used water quality data, user surveys, expert elicitation, and SEM to answer this research question.

An appealing strengthening of the researcher's approach for NCDWQ is the implementation of a user survey that collects information about the recreational uses and the user's perception of attainment of the water quality goals. The users have unique utilities that affect how they assess the eutrophication-related water quality goals, and, therefore, their responses can complement the information pertinent to the water quality data. The researchers are developing and testing a lake user survey that differs from previous approaches because the focus is the designated use translation and assessment of designated use attainment from people who directly use the specific waterbodies for recreation (fishing, swimming, boating, and aesthetics). In contrast, past surveys in Minnesota, Vermont, New York, and Florida used a nonrandom sample of volunteer water quality monitors and, as the authors of these papers acknowledge, their results may not be representative of the population of lake users (Smeltzer & Heiskary, 1990; Kishbaugh, 1993; Hoyer et al., 2004).

The user surveys were conducted concurrently with the summer 2005 North Carolina lake water quality sampling by NCDWQ. The surveys were designed, therefore, to learn about the lake users and how they assess the condition of the lake for their use(s) (see Appendix C). More specifically, the survey focuses on three areas of information:

1. the user's activities and familiarity with the lake,
2. the user's perception of the eutrophication-related water quality in relation to their use activities, and
3. demographic information.

It should be noted that the values of recreational lake users are usually expressed as categorical variables (ordinal or dichotomous scale), and, thus, some of the SEM statistical assumptions are violated.

The research also considers some of the existing correctives procedures from the SEM literature along with the development of model diagnostics and statistical tools for water quality data preparation (e.g., robust estimators to accommodate multivariate non-normality). As mentioned previously, this method is developed for lakes and reservoirs in North Carolina, but it can be easily transferred and employed by other states and assist the establishment of appropriate criteria and threshold levels for eutrophication-related designated use impairments.

4.3 Future Research Suggestions

One of the chief criticisms of the researchers' predictive nutrient criteria method is that they fail to suggest an appropriate threshold level for the predictive water quality variable. The reason that the researcher has not recommended a specific threshold level is that the criteria level should be set by a decision maker, who is the person who is best able to make the trade-off between environmental protection and TMDL research and implementation costs. Sometimes, however, people are not particularly good at making these tradeoffs. As a result, tools have been developed by decision analysts to assist decision makers. An important extension to the researchers' method of predictive nutrient criteria would be to assist decision makers in explicitly making this tradeoff. This could be accomplished by assessing the utility function of multiple decision makers and using this information to recommend the most appropriate water quality variable and threshold level.

The researchers would like to assess the utility function of the North Carolina Environmental Management Commission, the decision-making body for water quality standards in North Carolina. After assessing the utility functions for several decision makers, they would be able to analytically integrate the probability of use attainment, for a given criterion level, with the decision makers' utility function. The criterion level associated with the highest expected utility might then be chosen. Extending the currently funded research and assessing utility functions would provide a complete method for setting nutrient criteria. This alternative method could be developed into U.S. EPA guidance to provide states with an alternative method to set nutrient criteria that are predictively linked to the designated use.

4.4 Implications of This Research

The research team developed a method to set nutrient criteria that are predictively linked to the designated use. This project provides the foundation to develop alternative guidance on how states could choose appropriate nutrient criteria. It is believed that this regulatory choice is extremely important given the trade-off between environmental protection and TMDL research and cleanup. By providing decision makers with the nonbiased scientific information, they can explicitly make trade-offs that are inevitable when making any decision.

APPENDIX A

ORIGINAL EXPERT ELICITATION PROTOCOL

Dear INSERT NAME:

My name is INSERT and I am a graduate student at Duke University working with INSERT and Dr. Reckhow. I am writing to provide you with some background information on our project, and the expert elicitation process.

We are conducting a statistical modeling assessment intended to improve methods for the establishment of water quality standards. Specifically, we are addressing the U.S. EPA's national nutrient strategy, which requires states to establish eutrophication-related water quality criteria that are protective of designated uses.

We will be asking you a series of questions based on INSERT DESIGNATED USE for INSERT REGION. Your role in this process is to help identify a quantitative metric for the designated use statements. Our ultimate goal, following a phone interview, will be to estimate the INSERT DESIGNATED USE attainment/nonattainment threshold.

Included with this email (as attachments) are:

a) A few "pre-interview" questions. We will be using your responses to these questions to tailor our interview questions and use your time as efficiently as possible.

b) Statutory definitions of INSERT DESIGNATED USE, for your convenience.

Thank you for your time, and I look forward to speaking with you in the near future.

Sincerely,

INSERT NAME

Master of Environmental Management Candidate

Nicholas School of the Environment and Earth Sciences

Duke University, Durham, NC

INSERT EMAIL@duke.edu

Statement of Informed Consent

You have been invited to participate in an interview as a part of the research for INSERT NAME Master's Project at the Nicholas School of the Environment and Earth Sciences (advised by Kenneth H. Reckhow and Melissa A. Kenney). The goal of this study is to improve the understanding of the relationship between quantitative nutrient criteria and narrative water quality goal statements.

This interview will last approximately 1–2 hours. During the interview you will be asked to share your opinions on the appropriate quantitative metric for the narrative designated use statements of a waterbody. We will then use the information you provide during the interview to estimate points at which attainment/nonattainment of each designated use occur. There is no right or wrong answer to anything that will be discussed in the interview, and there is no certain answer that the researchers are hoping to hear from you. Results from this interview will be used to develop the following types of research reports: Master's Project, future research protocol, peer-reviewed manuscripts, grant and research reports, and presentations.

You were selected based on your knowledge of the waterbody system and your ability to provide insights regarding appropriate nutrient related water quality indicators. The data generated from the interview process will be recorded by written correspondence, digital recordings of phone and face-to-face communications, taking copious notes, and by creating formal elicitation reports. The results of the research may be presented in research reports. Access to data will be provided upon request after initial publication. Additionally, the information you provide us is not confidential and the interview and results that you provide us *may* be identified in research reports.

Your participation in this interview is entirely voluntary. You are also free to refuse to respond to any questions and to stop your participation at any time during the interview.

If you have any questions about this research project, please contact the project manager Melissa A. Kenney (m.kenney@duke.edu). If you have any questions about your rights as a research participant, please contact Ms. Lorna Hicks (lorna.hicks@duke.edu), Director of Human Protections Administration at Duke University.

I have read the information in this consent form and have been given the opportunity to discuss it and ask questions.

Print name Sign name Date

Preliminary Questions for INSERT REGION

CHECKLIST

INSERT REGION's Designated Uses:

select designated use that is most stringent to elicit judgments.

1. How would you define INSERT DESIGNATED USE?

    - Depends on the DU

2. What aspects of INSERT DESIGNATED USE do you consider to be most important?

    - Depends on the DU

3. If you had unlimited resources and were able to measure any biological, chemical, or physical variables, what variables would you measure to determine whether or not INSERT DESIGNATED USE was attained?

    - Higher level trophic indicators or other variable that makes sense?
    - Is there any other variables that would perfectly or near perfectly measure the designated use? If yes, what?
    - If multiple variables provided, what would be the single best variable?
    - Why would this variable be ideal to measure designated use?

4. Given the variable that you just identified as ideal to measure the aspects of INSERT DESIGNATED USE, what do you believe is the attainment vs. nonattainment change point level for this variable?

    - Measured value the same as #3.
    - Value provided?
    - Units clear?
    - Does it pass the clairvoyant test?

5. Given the variable that you just identified as ideal to measure the aspects of INSERT DESIGNATED USE, what commonly measured water quality variables would you use as a proxy for assessing your identified ideal variable?

    - N
    - P
    - Secchi depth
    - Temperature
    - pH
    - Chlorophyll *a*
    - What is the relationship among the variables you listed?
    - What is the ecological structure? (conceptual model)
    - Are there any seasonal effects that would change the model?

Expert Elicitation Phase 2

CHECKLIST

Question:

Given 100 lakes and reservoirs in INSERT REGION, all with identical average levels of these variables in the data row and assuming other factors not listed (e.g., morphological, climatic) vary randomly, how many of the 100 lakes and reservoirs would be in attainment of the designated use?

Example data rows:

Chla	DO	Secchi	TN	TP	DAP (#/L)	ZOOP (#/L)	....	# (out of 100) in attainment
1.8	9.2	4.1	201	4.6	0.69	11.78		?
1.4	9.2	5.5	187	6.2	0.70	21.62		?

The expert would answer the question that was stated above in the box labeled "# (out of 100) in attainment."

Procedure:

1. Look at a single data row. Look at all of the values for each of the variables. These variables were chosen based on your (the expert's) statement regarding measurable eutrophication related variables that are important for assessment of the designated use, using the ideal variable measure of designated use attainment.
2. Think about the relationship among the variables. Using this conceptual model, make an assessment of the number of lakes that will be in attainment of the designated use (out of 100 lakes).
3. Consider again the value you (the expert) chose as number of lakes in attainment of the designated use. If 5 more lakes had the exact same values for each of the variables in the data row, would the lake be in attainment of the designated use? If your answer changes, please make the appropriate modifications.
4. Once you are satisfied with the value chosen, move to the next data row and repeat the process.
5. When you have finished providing values for each of the data rows, please look through your responses to make sure that you still agree with your response. These values will be used as data in the structural equation models created for INSERT REGION.

APPENDIX B

REVISED EXPERT ELICITATION PROTOCOL

Dear INSERT,

I am a doctoral candidate in water quality modeling and decision analysis at the Duke University, Nicholas School of the Environment and Earth Sciences. For my dissertation (advised by Professors Ken Reckhow and Bob Clemen), I am developing a method for establishing nutrient criteria that are predictive of a waterbody's goals, or designated uses. Specifically, my research addresses the U.S. EPA's national nutrient strategy, which requires states to protect designated uses by establishing eutrophication-related water quality criteria. As a demonstration of our method, I am assisting North Carolina in re-assessing their nutrient standards for lakes and reservoirs.

Because there is no existing method for determining whether a waterbody is meeting its designated uses, such as swimmable or fishable, your judgments are needed to help us understand what characteristics of a lake lead to designated use attainment. Using your judgments, coupled with water quality data, we will be able to develop a model that links eutrophication and designated use attainment. (The model and this project do not try to assess compliance of a waterbody with the current water quality standards.)

For this project, I need to find individuals who are experienced and knowledgeable about North Carolina lakes. You were recommended as an expert by INSERT.

I am writing to find out whether you would be interested in serving as an expert for my study. The process includes two parts, both of which can be completed during a one-day meeting. The first part (approximately 2 hours) will include discussions about eutrophication processes and designated use impairment as well as the use of expert judgments in projects such as this one. In the second part (approximately 1 hour training; 2 hours on your own time), you will provide your judgments about designated use attainment. In this part of the study, I will provide you with data from a number of North Carolina lakes or reservoirs. For each case, you will have multiple measures of eutrophication-related water quality variables. Given these measures, you will be asked to assess the extent of eutrophication and designated use attainment for that particular lake. We will go through the first few cases together, so that you can work through the remaining cases on your own. In total, this process will take approximately 6 hours of your time.

After I receive your judgments, I will summarize them in a brief report, which I will send to you for your approval. You will be able at that time to modify or fine-tune your judgments as you see fit.

In return for your efforts, you will be acknowledged in the project's publications, and I will send you, if you wish, a copy of my dissertation and the resulting publications.

I hope that you will agree to participate, because I would enjoy working with you. Please contact me (919-613-8116; m.kenney@duke.edu) if you have any questions.

Sincerely,
Melissa Kenney

Expert Elicitation Discussion Questions for INSERT LAKE REGION

1. What are the mechanisms that lead to eutrophication?
    - Indicators
        - Nitrogen
        - Phosphorous
        - Chlorophyll *a*
        - Total suspended solids
        - Turbidity
        - Secchi Depth
        - DO
        - pH
        - Temperature
        - Benthic organism
    - Symptoms
        - Algal growth
        - Reduced clarity
        - Reduced oxygen levels
        - Fish kills
        - Ecosystem change
        - Excessive nutrient loading
    - Chain of events
        - Important factors or major effects of eutrophication in particular region

2. What other variables (noneutrophication) affect a waterbody's attainment of designated use?
    - Variables
        - Toxic algae
        - Sport fish population
        - Fecal coliform
    - Mechanisms
        - Land use
        - Pollutant loading
    - How do the other variables, mechanisms, etc., relate to eutrophication?
        - Fish biodiversity changing with trophic status
    - What does designated use mean? (if necessary)

Expert Elicitation

1. Consider a single data row. Look at all of the values for each of the variables.

2. Given the values for the specific variables in the data row:

Imagine 100 different lakes with the characteristics specified by the given data row. Of the 100 lakes, how many of the lakes would you expect to fall into each of the following five categories of eutrophication?

Category	Description
1	The lakes have: excellent water clarity, no color, very little algae, very low nutrient levels, very high oxygen, no odor, and very healthy, abundant aquatic life.
2	The lakes have: good water clarity, little color, little algae, low nutrient levels, high oxygen, little odor, and healthy, abundant aquatic life.
3	The lakes have: fair water clarity, some color, moderate amounts of algae, moderate nutrient levels, moderate oxygen, little odor, and somewhat healthy, abundant aquatic life.
4	The lakes have: poor water clarity, noticeable color, high algae, high nutrient levels, low oxygen, noticeable odor, and unhealthy, scarce aquatic life.
5	The lakes have: poor water clarity, considerable color, very high algae (likely scums), very high nutrient levels, low to no oxygen, strong offensive odor, and unhealthy, scarce aquatic life or no aquatic life.

(alternate presentation) The lake has:

	Water clarity	Color	Algae	Nutrient levels	Oxygen	Odor	Aquatic life
1	Excellent	None	Very little	Very low	Very high	No	Very healthy, abundant
2	Good	Little	Little	Low	High	Little	Healthy, abundant
3	Fair	Some	Moderate	Moderate	Moderate	Little	Somewhat healthy, abundant
4	Poor	Noticeable	High	High	Low	Noticeable	Unhealthy, scarce
5	Poor	Considerable	Very high	Very high	Low to no	Strong offensive	Unhealthy, scarce or none present

3. Given the values for the specific variables in the data row:

Imagine 100 different lakes with the characteristics specified by the given data row. Of the 100 lakes, how many of the lakes would you expect to fall into each of the five categories for the following designated uses?

A. Primary Contact Recreation (Swimming, Wakeboarding, Tubing, Water skiing, etc.)
   1. Excellent: Greatly exceeds expectations
   2. Very good: Exceeds expectations
   3. Acceptable: Meets expectations
   4. Fair: Below expectations
   5. Poor: Far below expectations

B. Secondary Contact Recreation (Fishing, Motorized Boating, Jet Skiing, Canoeing, Kayaking, etc.)
   1. Excellent: Greatly exceeds expectations
   2. Very good: Exceeds expectations
   3. Acceptable: Meets expectations
   4. Fair: Below expectations
   5. Poor: Far below expectations

C. Trout waters (include?)
   1. Excellent: Greatly exceeds expectations
   2. Very good: Exceeds expectations
   3. Acceptable: Meets expectations
   4. Fair: Below expectations
   5. Poor: Far below expectations

4. Once you are satisfied with the value chosen, move to the next data row and repeat the process.

5. When you have finished providing values for each of the data rows, please look through your responses to make sure that you still agree with your response. These values will be used as data in our study of eutrophication and designated use attainment in North Carolina INSERT LAKE REGION.

APPENDIX C

USER SURVEY PROTOCOL

June 4, 2005

Dear User Survey Participant:

You are invited to participate in a survey as a part of research for Melissa A. Kenney's doctoral dissertation project (http://www.duke.edu/mak22) at the Nicholas School of the Environment and Earth Sciences (advised by Kenneth H. Reckhow). The goal of this study is to better choose water quality indicators and levels that protect the use goals of North Carolina lakes.

You were selected because you are using the lake for one of its recreational goals. The survey will take approximately 3 minutes to complete. In the survey, you will be asked to share your opinions your use of the lake today. Your participation is entirely voluntary. You are free to refuse to respond to any questions and to stop your participation at any time during the survey. If you choose to complete the survey, we will use the survey data to help us assess whether the state is meeting its water quality goals. There is no right or wrong answer to your survey responses, and there is no particular answer that the researchers are hoping you will select.

Your survey responses are anonymous. Your name or contact information is not required or necessary for you to complete the survey. You may, however, provide your contact information on a separate form if you would like to receive a copy of the final report (available by September 2006) that will be generated using the survey data. The survey data will also be used to answer research questions that will lead to publishable manuscripts. You can keep up to date with the project progress at http://www.duke.edu/mak22.

If you have any questions about the survey, please contact Melissa A. Kenney (919-613-8133; m.kenney@duke.edu). If you have any questions about your rights as a research participant, please contact Ms. Lorna Hicks (lorna.hicks@duke.edu), Director of Human Protections Administration at Duke University.

Sincerely,

Melissa A. Kenney

Ph.D. Student in Water Quality Modeling and Decision Analysis

Duke University Nicholas School of the Environment and Earth Sciences

Enclosure (User Survey)

Lake User Survey

Please respond to the following questions for the conditions and your uses today.

1. What are you using the lake for today (check all that apply)?

    ____Swimming

    ____Fishing

    ____Lake Beauty/Tanning

    ____Picnicking / Hiking

    ____Jet skiing

    ____Canoeing / Kayaking

    ____Motorized boating

    ____Wakeboarding / Tubing / Water skiing

    ____Other (specify)____

2. How often do you use the lake?

    ____Couple times a week

    ____Once a week

    ____Twice a month

    ____Once a month

    ____Occasionally

    ____First time

3. What qualities of the water do you believe are important to be able to use the lake (check all that apply)?

    ____High clarity

    ____Water temperature

    ____Little water odor

    ____High oxygen level

    ____Little water color

    ____No algal scums

    ____Sport fish populations

    ____Little sediment (dirt) in the water

    ____Other (specify)____

4. The water of the lake is crystal clear today.

____Completely agree

____Agree

____Neutral

____Disagree

____Strongly disagree

5. What is the color of the water?

____ Blue

____ Green

____ Brown

____ Brownish / Green

____ Other (specify)______________________________

6. How would you describe the condition of the lake, in regards to the algae, today?

____The water is crystal clear. There are no apparent problems.

____It's not perfectly clear; there is a little algae.

____There is algal growth; water appears green, yellow, or brown from algae.

____There are high levels of algal growth limiting water clarity or causing odor.

____Severe water impacts from algal growth causing massive algal scums on the lake or washed up on shore, strong offensive odor, or fish kill.

____Other impacts, not algal related (specify)___________________________________

7. How would you describe the condition of the lake, in regards to the sediment, today?

____The water is crystal clear. There are no apparent problems.

____It's not perfectly clear; there is a little sediment.

____There is obvious sediment; water appears brown from sediment.

____There is a large amount of sediment limiting water clarity.

____Severe water impacts from sediment; the water appears and feels like muddy water.

____Other impacts, not sediment related (specify)___________________________________

8. The conditions of the lake are perfect for my use(s) today.

    ____Completely agree

    ____Agree

    ____Neutral

    ____Disagree

    ____Strongly Disagree

9. How would you rate the conditions of the lake based on your experiences today?

    a. Water quality

        ____Excellent

        ____Very good

        ____Acceptable

        ____Fair

        ____Poor

        ____N/A

    b. Lake Beauty

        ____Excellent

        ____Very good

        ____Acceptable

        ____Fair

        ____Poor

        ____N/A

    c. Swimming or Wakeboarding / Tubing / Water skiing

        ____Excellent

        ____Very good

        ____Acceptable

        ____Fair

        ____Poor

        ____N/A

d. Fishing

____Excellent

____Very good

____Acceptable

____Fair

____Poor

____N/A

e. Motorized Boating or Jet Skiing

____Excellent

____Very good

____Acceptable

____Fair

____Poor

____N/A

f. Canoeing or Kayaking

____Excellent

____Very good

____Acceptable

____Fair

____Poor

____N/A

10. What factors, if any, are impacting or limiting your use of the lake today (check all that apply)?

____Poor clarity

____Water temperature

____Water odor

____Low oxygen level

____Water color

____Algal scums

____Few sport fish populations

____Sediment (dirt) in the water

____None

11. If you went fishing, how many fish have you caught today?

____0

____1

____2

____3-4

____5-6

____more than 6

____N/A

12. If you went fishing, what type and how many of each fish have you caught today (if applicable)?

____Crappie

____Catfish

____Walleye

____Bass

____Bluegill

____Shad

____Sunfish

____Perch

____Trout

____I don't know

____N/A

____Other (specify)___________

Optional Demographic Information

A. What is your age?

____under 18

____18–25

____26–35

____36–49

____50–65

____65 or older

B. What is your gender?

____Male

____Female

C. What is your race or ethnicity?

____Caucasian/White

____African American

____American Indian

____Hispanic

____Pacific Islander

____Asian

____Other (specify)______________________

D. What is the highest level of education you have completed?

____some high school

____high school degree

____some college

____college degree

____some graduate school

____graduate degree

E. Where do you live?

____on lakeINSERT

____INSERT TOWN/CITY

____INSERT REGION

____North Carolina

____United States

____outside United States

F. Do you participate in an association related to the lake, such as homeowners' organization, friends of the lake, fishing club, etc.?

____No

____Yes (specify) ________________________________

APPENDIX D

EXPERT ELICITATION REFERENCES

This section was meant only to be a brief introduction to expert elicitation so that the reader understands the justification for using expert judgment and the protocol that the researchers developed for assessing those judgments. Given the nature of this report, it is impossible to provide a lengthy explanation of the expert elicitation theory. Therefore, if the reader is interested in understanding of expert elicitation, the following readings are recommended:

Clemen, R.T.; Reilly, T. *Making Hard Decisions with Decision Tools*. Duxbury Thomson Learning: Pacific Grove, Calif., 2001.

Keeney, R.L.; von Winterfeldt, D. A Prescriptive Risk Framework for Individual Health and Safety Decisions. *IEEE Trans. Eng. Manag.* **1991**, *38*, 191–201.

Meyer, M.A.; Booker, J.M. *Eliciting and Analyzing Expert Judgment: A Practical Guide*; Academic Press Ltd.: London, 1991.

Morgan, M.G.; Henrion, M. *Uncertainty: A Guide to Dealing with Uncertainty in Quantitative Risk and Policy Analysis*. Cambridge University Press: Cambridge, U.K., 1990.

APPENDIX E

Structural Equation Modeling Software and References

E.1 Structural Equation Modeling Software

To develop the SEM for this study, the research team used several SEM software packages and evaluated several others. The models included in this study were created using AMOS (Arbuckle, 2003) and Mplus (Muthen & Muthen, 2004). AMOS is an easy-to-use SEM package where the models are generated using a graphical interface. The modeler can then create SEM by drawing diagrams, which can be saved as JPEG images, and the program creates the equations and the covariance structure implied by the SEM diagram. Data can be easily input into the software package, however, it will only consider continuous variables. This package is excellent for individual learning SEM and would like to try out some of the ideas learned.

Another software package that the researchers used was Mplus. Mplus, unlike AMOS, does not include a graphical interface, but instead writing simple code creates the models. The benefit of Mplus is that the models can be easily modified and tested from one trial to the next. Additionally, Mplus is able to consider categorical and continuous variables.

E.2 Structural Equation Modeling References

This section was meant only to be a brief introduction to SEM so that one can have a conceptual understanding of the quantitative technique that was used to test the hypothesis and generate conclusions. Given the nature of this report, it is impossible to provide a lengthy explanation of the statistical details of SEM. Therefore, for anyone interested in gaining a deeper understanding of SEM to apply the researcher's method, the following readings are recommended:

Shipley, B. *Cause and Correlation in Biology: A User's Guide to Path Analysis, Structural Equations and Causal Inference.* Cambridge University Press: Cambridge, UK, 2000.

Bollen, K.A. *Structural Equations with Latent Variables.* Wiley and Sons: New York, 1989.

McCune, B.; Grace, J.B. *Analysis of Ecological Communities.* MjM Software Design, Oregon, 2002.

Kline, R.B. *Principles and Practice of Structural Equation Modeling.* The Guilford Press: New York, 1998.

Jöreskog, K.G. A General Method for Estimating a Linear Structural Equation System. In *Structural Equation Model in the Social Sciences*; Goldberger, A.S.; Duncan, O.D., eds.; Academic Press: New York, 1973; pp 85–112.

Jöreskog, K.G., Sörbom, D. *LISREL 8: Structural Equation Modeling with the SIMPLIS Command Language.* Erlbaum Associates: Hillsdale, NJ, 1993.

Muthen, L.K.; Muthen, B.O. *Mplus Statistical Analysis with Latent Variables*. UCLA: Los Angeles, 2004.

References

American Rivers Foundation. *Report on the 20 Most Threatened American Rivers.* American Rivers Foundation: New York, 1997.

Arbuckle, J.L. Amos (software). SmallWaters Corporation: Chicago, 2003.

Arhonditsis, G.; Brett, M.T.; Frodge, J. Environmental Control and Limnological Impacts of a Large Recurrent Spring Bloom in Lake Washington, USA. *Environ. Manag.* **2003**, *31*, 603–618.

Arhonditsis, G.B.; Winter, M.; Brett, M.T.; Schindler, D.E. Patterns and Mechanisms of Phytoplankton Variability in Lake Washington (USA). *Water Res.* **2004**, 38, 4013-4027.

Arhonditsis, G.B.; Paerl, H.W.; Valdes, L.M.; Stow, C.A.; Steinberg, L.J.; Reckhow, K.H. Application of Bayesian Structural Equation Modeling for Exploring the Neuse River Estuary (NC, USA) Phytoplankton Dynamics. *Limnol. Oceanogr.,* submitted.

Arhonditsis, G.B.; Stow, C.A.; Paerl, H.W.; Valdes, L.M.; Steinberg, L.J.; Reckhow, K.H. Delineation of the Role of Nutrient Dynamics and Hydrologic Forcing on the Neuse River Estuary (NC, USA) Spatiotemporal Phytoplankton Patterns. *Ecosystems,* submitted.

Arhonditsis, G.B.; Stow, C.A.; Steinberg, L.J.; Kenney, M.A.; Lathrop, R.C.; McBride, S.J.; Reckhow. K.H. Exploring Ecological Patterns with Structural Equation Modelling and Bayesian Analysis.. *Ecol. Model.* **2006** *192*, 385–409.

Barnett, V.; O'Hagan, A. *Setting Environmental Standards: The Statistical Approach to Handling Uncertainty and Variation*; Chapman & Hall/CRC: New York, 1997.

Bentler, P.M. Comparative Fit Indexes in Structural Models. *Psychol. Bull.* **1990**, *107*, 238–246.

Bettman, J.R.; Luce, M.F.; Payne, J.W. Constructive Consumer Choice Processes. *J. Consum. Res.* **1998**, *25*, 187–217.

Bollen, K.A. *Structural Equations with Latent Variables.* Wiley and Sons: New York, 1989.

Borsuk, M.E. Predictive Assessment of Fish Health and Fish Kills in the Neuse River Estuary Using Elicited Expert Judgment. *Hum. Ecol. Risk Assess.* **2004**, *10*, 415–434.

Borsuk, M.E.; Higdon, D.; Stow, C.A.; Reckhow, K.H. A Bayesian Hierarchical Model to Predict Benthic Oxygen Demand from Organic Matter Loading in Estuaries and Coastal Zones. *Ecol. Model.* **2001**, *143*, 165–181.

Borsuk, M.E.; Powers, S.P.; Peterson, C.H. A Survival Model of the Effects of Bottom-water Hypoxia on the Population Density of an Estuarine Clam (*Macoma balthica*). *Can. J. Fish Aquat. Sci.* **2002**, *59,* 1266–1274.

Borsuk, M.E.; Stow, C.A.; Reckhow, K.H. Confounding Effect of Flow on Estuarine Response to Nitrogen Loading. *J. Environ. Eng. ASCE* **2004**, *130*, 605–614.

Clemen, R.T., Reilly., T. *Making Hard Decisions with Decision Tools.* Duxbury Thomson Learning: Pacific Grove, Calif., 2001.

Cloern, J.E. Our Evolving Conceptual Model of the Coastal Eutrophication Problem. *Mar. Ecol.-Prog. Ser.* **2001**, *210*, 223–253.

Congdon, P. *Applied Bayesian Modelling*. Wiley Series in Probability and Statistics: West Sussex, England, 2003.

Coupey, E.; Irwin, J.R.; Payne, J.W. Product Category Familiarity and Preference Construction. *J. Consum. Res.* **1998**, *24*, 285–294.

Edmondson, W.T. Sixty Years of Lake Washington: A Curriculum Vitae. *Lake Reserv. Manage.* **1994**, *10*, 75–84.

Grace, J.B.; Pugesek, B.H. A Structural Equation Model of Plant Species Richness and Its Application to a Coastal Wetland. *Am. Nat.* **1997**, *3*, 436–460.

Grace, J.B.; Pugesek, B.H. On the Use of Path Analysis and Related Procedures for the Investigation of Ecological Problems. *Am. Nat.* **1998**, *152*, 151–159.

Hampel, F.R.; Ronchetti, E.M.; Rousseeuw, P.J.; Stahel, W.A. *Robust Statistics: The Approach Based on Influenced Function.* John Wiley & Sons: New York. 1986.

Hoelter, J.W. The Analysis of Covariance Structures: Goodness-of-Fit Indices. *Sociol. Method. Res.* **1983**, *11*, 325–344.

Hoyer, M., Brown, C., Canfield. D. Relations between Water Chemistry and Water Quality as Defined by Lake Users in Florida. *Lake Reserv. Manage.* **2004**, *20*, 240–248.

Jöreskog, K.G. A General Method for Estimating a Linear Structural Equation System. In *Structural Equation Model in the Social Sciences*; Goldberger, A.S.; Duncan, O.D., eds.; Academic Press: New York, 1973; pp 85–112.

Jöreskog, K.G., Sörbom, D. *LISREL 8: Structural Equation Modeling with the SIMPLIS Command Language.* Erlbaum Associates: Hillsdale, NJ, 1993.

Kass, R.E.; Raftery, A.E. Bayes Factors. *J. Am. Stat. Assoc.* **1995**, *90*, 773–795.

Keeney, R.L.; von Winterfeldt, D. A Prescriptive Risk Framework for Individual Health and Safety Decisions. *IEEE Trans. Eng. Manage.* **1991**, *38*, 191–201.

Kenney, M.A. A Predictive Approach to Selecting Eutrophication-Related Criteria and Threshold Levels for Lakes and Reservoirs in Western North Carolina; Duke University, Durham, N.C., 2005. (Copy available from m.kenney@duke.edu.)

Kerwin, J. *Salmon and Steelhead Habitat Limiting Factors Report for the Cedar-Sammamish Basin (Water Resource Inventory Area 8).* Washington Conservation Commission: Olympia, Wash., 2001.

Kishbaugh, S. *Applications and Limitations of Qualitative Lake Assessment Data.* North American Lake Management Society: Seattle, Wash., 1993.

Kline, R.B. *Principles and Practice of Structural Equation Modeling.* The Guilford Press: New York, 1998.

Lathrop, R.C.; Carpenter, S.R.; Robertson, D.M. Summer Water Clarity Responses to Phosphorus, *Daphnia* Grazing, and Internal Mixing in Lake Mendota. *Limnol. Oceanogr.* **1999**, *44*, 137–146.

Lathrop, R.C.; Carpenter, S.R.; Rudstam, L.G. Water Clarity in Lake Mendota since 1900: Responses to Differing Levels of Nutrients and Herbivory. *Can. J. Fish Aquat. Sci.* **1996**, *53*, 2250–2261.

Lehman, J.T. Hypolimnetic Metabolism in Lake Washington – Relative Effects of Nutrient Load and Food Web Structure on Lake Productivity. *Limnol. Oceanogr.* **1988**, *33*, 1334–1347.

Malaeb, Z.A.; Summers, J.K.; Pugesek, B.H. Using Structural Equation Modeling to Investigate Relationships among Ecological Variables. *Environ. Ecol. Stat.* **2000** *7*, 93–111.

McCune, B.; Grace, J.B. *Analysis of Ecological Communities.* MjM Software Design, Ore., 2002.

Meyer, M.A.; Booker, J.M. *Eliciting and Analyzing Expert Judgment: A Practical Guide*; Academic Press, Ltd.: London, 1991.

Morgan, M.G.; Henrion, M. *Uncertainty: A Guide to Dealing with Uncertainty in Quantitative Risk and Policy Analysis*; Cambridge University Press: Cambridge, U.K., 1990.

Muthen, L.K.; Muthen, B.O. *Mplus Statistical Analysis with Latent Variables.* UCLA: Los Angeles, 2004.

North Carolina Department of Water Quality. *The Neuse Nutrient Sensitive Waters Management Strategy.* North Carolina Department of Water Quality: Raleigh, N.C., 1997.

North Carolina Department of Water Quality. *North Carolina Administrative Code. Surface Waters and Wetland Standards.* North Carolina Department of Water Quality: Raleigh, N.C., 2003a.

North Carolina Department of Water Quality. *Surface Water Classifications*, 2003b. Available at http://h2o.enr.state.nc.us/csu/swc.html.

Pace, M.L. Prediction and the Aquatic Sciences. *Can. J. Fish Aquat. Sci.* **2001** *58*, 63–72.

Paerl, H.W.; Valdes, L.M.; Adolf, J.; Peierls B.L.; Harding, Jr., L.W. Anthropogenic and Climatic Influences on the Eutrophication of Large Estuarine Ecosystems. *Limnol Oceanogr.* **2006**, *51*(1, part 2), 448–462.

Paerl, H.W.; Valdes, L.M.; Joyner, A.R.; Piehler, M.F. Solving Problems Resulting from Solutions: Evolution of a Dual Nutrient Management Strategy for the Eutrophying Neuse River Estuary, North Carolina. *Environ. Sci. Technol.* **2004**, *38*, 3068–3073.

Paerl, H.W.; Valdes, L.M.; Pinckney, J.L.; Piehler, M.F.; Dyble, J.; Moisander, P.H. Phytoplankton Photopigments as Indicators of Estuarine and Coastal Eutrophication. *BioScience* **2003**, *53*, 953–964.

Payne, J.W.; Bettman, J.R.; Schkade, D.A. Measuring Constructed Preferences: Towards a Building Code. *J. Risk Uncertainty* **1999,** *19,* 243–270.

Pinckney, J.L.; Millie, D.F.; Vinyard, B.T.; Paerl, H.W. Environmental Controls of Phytoplankton Bloom Dynamics in the Neuse River Estuary, North Carolina, USA. *Can. J. Fish Aquat. Sci.* **1997,** *54*, 2491–2501.

Press, J.S.; Tanur, J.M. *The Subjectivity of Science and the Bayesian Approach.* John Wiley & Sons: New York, 2001.

Pugesek, B.H.; Grace, J.B. On the Utility of Path Modelling for Ecological and Evolutionary Studies. *Funct. Ecol.* **1998**, *5*, 853–856.

Qian, S.S.; Borsuk, M.E.; Stow, C.A. Seasonal and Long-term Nutrient Trend Decomposition along a Spatial Gradient in the Neuse River Watershed. *Environ. Sci. Technol.* **2000**, *34*, 4474–4482.

Reckhow, K.H. A Comparison of Robust Bayes and Classical Estimators for Regional Lake Models of Fish Response to Acidification. *Water Resour. Res.* **1988**, *24*, 1061–1068.

Reckhow, K.H.; Arhonditsis, G.B.; Kenney, M.A.; Hauser, L.; Tribo, J.; Wu, C.; Elcock, K.J.; Steinberg, L.J.; Stow, C.A.; McBride S.J. A Predictive Approach to Nutrient Criteria. *Environ. Sci. Technol.* **2005,** *39*, 2913–2919.

Redfield, A.C. The Biological Control of the Chemical Factors in the Environment. *Am. Sci.* **1958,** *46*, 205–230.

Scheines, R.; Hoijtink, H.; Boomsma, A. Bayesian Estimation and Testing of Structural Equation Models. *Psychometrika* **1999**, *64*, 37–52.

Shipley, B. Exploratory Path Analysis with Applications in Ecology and Evolution. *Am. Nat.* **1997,** *149*, 1113–1138.

Shipley, B. *Cause and Correlation in Biology: A User's Guide to Path Analysis, Structural Equations and Causal Inference.* Cambridge University Press: Cambridge U.K., 2000.

Smeltzer, E., Heiskary, S. Analysis and Application of Lake User Survey Data. *Lake Reserv. Manage.* **1990,** *6*, 109–118.

Smith, F.A. Scaling of Digestive Efficiency with Body-Mass in *Neotoma*. *Funct. Ecol.* **1995,** *9*, 299–305.

Smith, F. A.; Brown, J. H.; Valone, T. J. Path Modeling Methods and Ecological Interactions: A Response to Grace and Pugesek. *Am. Nat.* **1998**, *152* (1), 160–161.

Sommer, U. The Role of Competition for Resources in Phytoplankton Succession. In *Phytoplankton Ecology. Succession in Plankton Communities;* Sommer, U., ed.; Springer Verlag: New York, 1989; pp 57–106.

Soranno, P.A. Factors Affecting the Timing of Surface Scums and Epilimnetic Blooms of Blue-green Algae in a Eutrophic Lake. *Can. J. Fish Aquat. Sci.* **1997**, *54*, 1965–1975.

Stow, C.A.; Borsuk, M.E., 2003. Enhancing Causal Assessment of Estuarine Fishkills Using Graphical Models. *Ecosystems* **2003**, *6*, 11–19.

Twomey, L.; John, J.; Thompson, P. Seasonal Succession of Diatoms and Other Phytoplankton in a Bar Built Estuary, Wilson Inlet, Western Australia. In *Proceedings of the 15th International Diatom Symposium*; John, J., ed.; Königstein, Koeltz Scientific Books: Perth, Australia, 2002; p 395–420.

U.S. Environmental Protection Agency. *Clean Water Action Plan: Restoring and Protecting America's Waters;* EPA-840-R-98-001; U.S. Environmental Protection Agency: Washington, D.C., 1988.

U.S. Environmental Protection Agency. *Atlas of America's Polluted Waters;* EPA 840-B00-002. Office of Water (4503F); U.S. Environmental Protection Agency: Washington, D.C., 2000a.

U.S. Environmental Protection Agency. *Ambient Water Quality Criteria Recommendations: Information Supporting the Development of State and Tribal Nutrient Criteria for Lakes and Reservoirs in Nutrient Ecoregion IX*; EPA 822-B-00-011; Office of Water, U.S. Environmental Protection Agency: Washington, D.C., 2000b.

U.S. Environmental Protection Agency. *Ambient Water Quality Criteria Recommendations: Information Supporting the Development of State and Tribal Nutrient Criteria for Lakes and Reservoirs in Nutrient Ecoregion XIV*; EPA 822-B-01-011; U.S. Environmental Protection Agency: Washington, D.C., 2001.

Washington State Department of Ecology. *Water Quality Standards for Surface Waters of the State of Washington;* Chap. 173-201A WAC; Washington State Department of Ecology: Olympia, Wash., 2003.

Weiher, E.; Forbes, S.; Schauwecker, T.; Grace, J.B. Multivariate Control of Plant Species Richness and Community Biomass in Blackland Prairie. *Oikos* **2004**, *106*(1), 151–157.

WASTEWATER UTILITY

Alabama
Montgomery Water Works & Sanitary Sewer Board

Alaska
Anchorage Water & Wastewater Utility

Arizona
Gila Resources
Glendale, City of, Utilities Department
Mesa, City of
Peoria, City of
Phoenix Water Services Department
Pima County Wastewater Management

Arkansas
Little Rock Wastewater Utility

California
Calaveras County Water District
Central Contra Costa Sanitary District
Corona, City of
Crestline Sanitation District
Delta Diablo Sanitation District
Dublin San Ramon Services District
East Bay Dischargers Authority
East Bay Municipal Utility District
Eastern Municipal Water District
El Dorado Irrigation District
Fairfield-Suisun Sewer District
Fresno Department of Public Utilities
Inland Empire Utilities Agency
Irvine Ranch Water District
Las Virgenes Municipal Water District
Livermore, City of
Lodi, City of
Los Angeles, City of
Napa Sanitation District
Orange County Sanitation District
Palo Alto, City of
Riverside, City of
Sacramento Regional County Sanitation District
San Diego Metropolitan Wastewater Department, City of
San Francisco, City & County of
Sanitation Districts of Los Angeles County
San Jose, City of
Santa Barbara, City of
Santa Cruz, City of
Santa Rosa, City of
South Bayside System Authority
South Coast Water District
South Orange County Wastewater Authority
Stege Sanitary District
Sunnyvale, City of
Union Sanitary District
West Valley Sanitation District

Colorado
Aurora, City of
Boulder, City of
Colorado Springs Utilities
Greeley, City of
Littleton/Englewood Water Pollution Control Plant
Metro Wastewater Reclamation District, Denver

Connecticut
The Mattabassett District
Greater New Haven WPCA

District of Columbia
District of Columbia Water & Sewer Authority

Florida
Broward, County of
Fort Lauderdale, City of
JEA
Miami-Dade Water & Sewer Authority
Orange County Utilities Department
Reedy Creek Improvement District
Seminole County Environmental Services
St. Petersburg, City of
Stuart Public Utilities
Tallahassee, City of
Tampa, City of
Toho Water Authority
West Palm Beach, City of

Georgia
Atlanta Department of Watershed Management
Augusta, City of
Clayton County Water Authority
Cobb County Water System
Columbus Water Works
Fulton County
Gwinnett County Department of Public Utilities
Savannah, City of

Hawaii
Honolulu, City & County of

Idaho
Boise, City of

Illinois
American Bottoms Wastewater Treatment Plant
Greater Peoria Sanitary District
Kankakee River Metropolitan Agency
Metropolitan Water Reclamation District of Greater Chicago
Wheaton Sanitary District

Iowa
Ames, City of
Cedar Rapids Wastewater Facility
Des Moines, City of
Iowa City

Kansas
Johnson County Unified Wastewater Districts
Unified Government of Wyandotte County/ Kansas City, City of

Kentucky
Louisville & Jefferson County Metropolitan Sewer District

Louisiana
Sewerage & Water Board of New Orleans

Maine
Bangor, City of
Portland Water District

Maryland
Anne Arundel County Bureau of Utility Operations
Hagerstown, City of
Howard County Department of Public Works
Washington Suburban Sanitary Commission

Massachusetts
Boston Water & Sewer Commission
Upper Blackstone Water Pollution Abatement District

Michigan
Ann Arbor, City of
Detroit, City of
Holland Board of Public Works
Lansing, City of
Saginaw, City of
Wayne County Department of Environment
Wyoming, City of

Minnesota
Rochester, City of
Western Lake Superior Sanitary District

Missouri
Independence, City of
Kansas City Missouri Water Services Department
Little Blue Valley Sewer District
Metropolitan St. Louis Sewer District

Nebraska
Lincoln Wastewater System

Nevada
Henderson, City of
Reno, City of

New Jersey
Bergen County Utilities Authority
Ocean, County of
Passaic Valley Sewerage Commissioners

New York
New York City Department of Environmental Protection

North Carolina
Charlotte/Mecklenburg Utilities
Durham, City of
Metropolitan Sewerage District of Buncombe County
Orange Water & Sewer Authority

Ohio
Akron, City of
Butler County Department of Environmental Services
Columbus, City of
Metropolitan Sewer District of Greater Cincinnati
Northeast Ohio Regional Sewer District
Summit, County of

Oklahoma
Tulsa, City of

Oregon
Clean Water Services
Eugene/Springfield Water Pollution Control
Water Environment Services

Pennsylvania
Philadelphia, City of
University Area Joint Authority

South Carolina
Charleston Commissioners of Public Works
Mount Pleasant Waterworks & Sewer Commission
Spartanburg Sanitary Sewer District

Tennessee
Cleveland, City of
Knoxville Utilities Board
Murfreesboro Water & Sewer Department
Nashville Metro Water Services

Texas
Austin, City of
Dallas Water Utilities
Denton, City of
El Paso Water Utilities
Fort Worth, City of
Gulf Coast Waste Disposal Authority
Houston, City of
San Antonio Water System
Trinity River Authority

Utah
Salt Lake City Corporation

Virginia
Alexandria Sanitation Authority

WERF SUBSCRIBERS

Arlington, County of
Fairfax County
Hampton Roads Sanitation District
Henrico, County of
Hopewell Regional Wastewater Treatment Facility
Loudoun County Sanitation Authority
Lynchburg Regional WWTP
Prince William County Service Authority
Richmond, City of
Rivanna Water & Sewer Authority

Washington
Everett, City of
King County Department of Natural Resources
Seattle Public Utilities
Sunnyside, Port of
Yakima, City of

Wisconsin
Green Bay Metro Sewerage District
Kenosha Water Utility
Madison Metropolitan Sewerage District
Milwaukee Metropolitan Sewerage District
Racine, City of
Sheboygan Regional Wastewater Treatment
Wausau Water Works

Australia
South Australian Water Corporation
Sydney Water Corporation
Water Corporation of Western Australia

Canada
Greater Vancouver Regional District
Regina, City of, Saskatchewan
Toronto, City of, Ontario
Winnipeg, City of, Manitoba

New Zealand
Watercare Services Limited

United Kingdom
Yorkshire Water Services Limited

STORMWATER UTILITY

California
Los Angeles, City of, Department of Public Works
Monterey, City of
San Francisco, City & County of
Santa Rosa, City of

Colorado
Aurora, City of
Boulder, City of

Georgia
Griffin, City of

Iowa
Cedar Rapids Wastewater Facility
Des Moines, City of

Kansas
Overland Park, City of

Maine
Portland Water District

Minnesota
Western Lake Superior Sanitary District

North Carolina
Charlotte, City of, Stormwater Services

Pennsylvania
Philadelphia, City of

Tennessee
Chattanooga Stormwater Management

Washington
Bellevue Utilities Department
Seattle Public Utilities

STATE

Arkansas Department of Environmental Quality
Fresno Metropolitan Flood Control District
Kansas Department of Health & Environment
Kentucky Department of Environmental Protection
Ohio River Valley Sanitation Commission
Urban Drainage & Flood Control District, CO

CORPORATE

Abt. Associates
ADS Environmental Services
Alan Plummer & Associates
Alden Research Laboratory Inc.
American Water
Aqua-Aerobic Systems Inc.
Aquateam–Norwegian Water Technology Centre A/S
ARCADIS
Associated Engineering
Black & Veatch
Blasland, Bouck & Lee Inc.
Boyle Engineering Corporation
Brown & Caldwell
Burns & McDonnell
CABE Associates Inc.
The Cadmus Group
Camp Dresser & McKee Inc.
Carollo Engineers Inc.
Carpenter Environmental Associates Inc.
CDS Technologies Inc.
Chemtrac Systems Inc.
CH2M HILL
Construction & Environmental Consulting, Inc.
D&B/Guarino Engineers, LLC
Damon S. Williams Associates, LLC
Dewling Associates Inc.
Earth Tech Inc.
Ecovation
EMA Inc.
Environ/The ADVENT Group, Inc.
The Eshelman Company Inc.
EWT Holdings Corporation
Fay, Spofford, & Thorndike Inc.
Freese & Nichols Inc.
ftn Associates Inc.
Fuss & O'Neill Inc.
Gannett Fleming Inc.
Geosyntec Consultants
GHD
Golder Associates Ltd.
Greeley and Hansen LLC
Hazen & Sawyer, P.C.
HDR Engineering Inc.
HNTB Corporation
HydroQual Inc.
Infilco Degremont Inc.
Jacobson Helgoth Consultants Inc.
Jason Consultants LLC Inc.
Jordan, Jones, & Goulding Inc.
KCI Technologies Inc.
Kelly & Weaver, P.C.
Kennedy/Jenks Consultants
KMK Consultants
Komline Sanderson Engineering Corporation
Lawler, Matusky & Skelly Engineers, LLP
Limno-Tech Inc.
Lombardo Associates Inc.
Malcolm Pirnie Inc.
Material Matters
McKim & Creed
MEC Analytical Systems Inc.
Metcalf & Eddy Inc.
MPR Engineering Corporation, Inc.
MWH
New England Organics
O'Brien & Gere Engineers Inc.
Odor & Corrosion Technology Consultants Inc.
Oswald Green, LLC
PA Government Services Inc.
Parametrix Inc.
Parsons
Post, Buckley, Schuh & Jernigan
R&D Engineering/Conestoga Rover & Associates
RMC Inc.
R.M. Towill Corporation
Ross & Associates Ltd.
Rothberg, Tamburini & Windsor, Inc.
Royce Technologies
Stantec Consulting Inc.
Stearns & Wheler, LLC
Stone Environmental Inc.
Stormwater360
Stratus Consulting Inc.
Synagro Technologies Inc.
Tetra Tech Inc.
Trojan Technologies Inc.
Trussell Technologies, Inc.
URS Corporation
USFilter
Wade-Trim Inc.
Westin Engineering Inc.
Weston Solutions Inc.
Woodard & Curran
WRc/D&B, LLC
WWETCO, LLC
Zenon Environmental Inc.
Zoeller Pump Company

INDUSTRY

American Electric Power
ChevronTexaco Energy Research & Technology Company
The Coca-Cola Company
Dow Chemical Company
DuPont Company
Eastman Chemical Company
Eastman Kodak Company
Eli Lilly & Company
Merck & Company Inc.
Premier Chemicals LLC
Procter & Gamble Company
RWE Thames Water Plc
Severn Trent Services Inc.
Suez Environnment
United Water Services LLC

Note: List as of 4/1/06

IWA Publishing's authorised EU representative for General Product Safety Regulations is Diane D'Arras, 15 rue Duret, 75116 Paris, France, e-mail: safety@iwap.co.uk.

Printed and bound by CPI Group (UK) Ltd, Croydon, CR0 4YY
12/05/2026
02108892-0006